**Alexandre Antunes Brum**
**Aline Neutzling Brum**

# Substrate degradation by sun mushroom

**Alexandre Antunes Brum**
**Aline Neutzling Brum**

# Substrate degradation by sun mushroom

## States of lignocellulose

**Imprint**

Any brand names and product names mentioned in this book are subject to trademark, brand or patent protection and are trademarks or registered trademarks of their respective holders. The use of brand names, product names, common names, trade names, product descriptions etc. even without a particular marking in this work is in no way to be construed to mean that such names may be regarded as unrestricted in respect of trademark and brand protection legislation and could thus be used by anyone.

Cover image: www.ingimage.com

This book is a translation from the original published under ISBN 978-613-9-65050-7.

Publisher:
Sciencia Scripts
is a trademark of
Dodo Books Indian Ocean Ltd. and OmniScriptum S.R.L publishing group

120 High Road, East Finchley, London, N2 9ED, United Kingdom
Str. Armeneasca 28/1, office 1, Chisinau MD-2012, Republic of Moldova, Europe
Printed at: see last page
**ISBN: 978-620-7-85033-4**

# Contents

# CHAPTER 1

## 1. INTRODUCTION

Fungi are one of the most important groups of microorganisms in the decomposition of organic matter due to their specialised degradation capacity. This activity occurs above all through their vegetative or mycelial phase.

In the vegetative and reproductive phases, biomass formation depends on the production of extracellular enzymes, which are fundamental in the degradation of substrate components, mainly lignocellulose (VELÁZQUEZ-CEDENO et al. 2002).

Basically, there are two groups of enzymes responsible for the degradation of lignocellulose: oxidative enzymes and hydrolytic enzymes. Oxidative enzymes, such as laccase, manganese peroxidase and lignin peroxidase, work to degrade lignin and detoxify the growth medium of the metabolites generated during degradation. The degradation and/or biotransformation of lignin allows hydrolytic enzymes such as cellulase, |3-glucosidase, xylanase and others to act on carbon sources, enabling the mycelium to absorb polysaccharides, which are a source of carbon mainly for the formation of fruiting (DURRANT et al. 1991).

In some groups of fungi, during the reproductive phase, fruiting structures are formed, called mushrooms, which depend on the mycelial biomass to supply them with water and nutrients (MOL, 1989). The nutritional and environmental requirements of the reproductive phase differ from those that are optimal for mycelial growth and, therefore, the study of the ecology and physiology of a fungus must consider both phases of the cycle.

vegetative capacity, among others, which have a direct impact on productivity.

Considering that the productivity of a crop depends on three basic factors: the strain of fungus used, the environmental conditions provided at the different stages of cultivation and the quality of the substrate, it is essential to optimise each of these factors. On the other hand, an increasing number of growers in the state of Santa Catarina are establishing non-seasonal crops in grow houses, using different levels of environmental control technology. On the other hand, studies into aspects of the substrate are quite restricted and little information is available, especially on the use of new waste products in the formulation of compost and technologies for preparing this substrate.

The compost used to grow A. brasiliensis is a substrate that has undergone an essentially aerobic degradation process and is then pasteurised, initially based on the technology developed for growing champignons, Agaricus brunnescens Peck. [= Agaricus bisporus Lange (Imbach)], both in terms of inputs for formulations and methodologies for managing composting yards and pasteurisation tunnels. However, the ecological and nutritional requirements of A. brasiliensis are different to those required by A. brunnescens, both in the vegetative and reproductive phases, which implies the occurrence of diseases and low yields, as reported by Nascimento & Eira (2003) where the incidence of false truffle (Diehliomyces microsporus Gilkey) in the states of São Paulo and Paraná, started to occur due to inadequate pasteurisation conditions for the compost.

In order to produce a selective compound for A. brasiliensis, it is important to know the physiological characteristics of the species, particularly its degradative capacity in relation to the main components of the substrate, lignocellulosic materials.

There is no scientific information on the enzyme profile, its degradative capacity, as well as the dynamics of lignocellulose degradation by A. brasiliensis, and its versatility to grow and fruit on different agro-industrial residues.

The study of alternative cultivation technologies that optimise the heat treatment of the substrate in particular could be an important strategy for increasing the productivity of A. brasiliensis cultivations. In addition, the inclusion of new ingredients in formulations, based on studies of the degradative capacity of A. brasiliensis, is also important due to the regional predominance of residues.

Production in sterilised, non-composted substrates is an alternative that can make the use of regionally predominant waste viable, as well as allowing the use of infrastructure identical to that of other species such as shiitake [Lentinula edodes (Berk.) Pegler] (CAMPBELL & SLEE, 1987).

Therefore, before developing cultivation technologies, especially those relating to the preparation of cultivation substrates, it is essential to study the physiological characteristics of A. brasiliensis, particularly its enzyme profile and the dynamics of enzyme activity during growth on lignocellulosic waste, as well as the dynamics of the degradation process.

## 2.  OBJECTIVES

**2.1**  General objective

- To study the dynamics of lignocellulosic degradation by Agaricus brasiliensis, using biochemical and spectroscopic methods, during the vegetative growth of Agaricus brasiliensis in two cultivation systems, axenic and traditional;


**2.2**  Specific objectives

-    Determine the lignocellulolytic enzyme profile of A. brasiliensis;

-    Monitor the dynamics of the enzymatic activity of A. brasiliensis;

-    Evaluate the production of vegetative biomass and identify possible indirect markers of vegetative growth;

-    Detecting chemical changes in the different substrates;


## 3.  LITERATURE REVIEW

**3.1**  Agaricus brasiliensis or Agaricus blazei (Murrill) ss. Heinemann

Agaricus brasiliensis S. Wasser & Didukh (Agaricomycetideae), and more recently A. subrufescens, is the current name for the fungus colloquially known as mushroom of the sun®, piedade mushroom, almond portobello, kawahiratake, himematsutake, champignon do Brasil and others. Until recently, Agaricus blazei (Murrill) ss. Heinemann, a fungus native to the Americas whose popularity was based on the β-glucans present in its cell wall, polysaccharides with bioactivity, mainly immunomodulatory (TAKATU, 2001; WASSER et al. 2002). In 1993, using specimens of A. blazei (Murrill) native to Brazil, researcher Heinemann noted chemical-morphological differences in the fruiting of this fungus compared to specimens from North America, and renamed them A. blazei (Murrill) ss. Heinemann. In 2002, Wasser et al. proposed the name A. brasiliensis for the Brazilian specimens.

**3.2**  Vegetative growth


The study of the vegetative or mycelial growth of fungi, whose interest lies in the formation of fruiting bodies, mushrooms, is fundamental, since the development of these fruiting bodies is closely linked to the mycelium (MOL, 1989).

The ability of a fungus to produce fungistatic and bacteriostatic substances, as well as tolerate them; to have high growth rates when stimulated by soluble nutrients; to produce appropriate and efficient enzymes for the degradation of recalcitrant constituents (GARRET, 1970

apud SAVOIE et al. 1996), can determine the efficient colonisation of a substrate by a particular species or strain (SAVOIE et al. 1996).

Carbon utilisation is fundamental in the construction and development of hyphae, which grow apically as a result of the deposition of new material, mainly during the vegetative phase. Under suitable environmental conditions, such as low $CO_2$ levels and suitable temperatures, the hyphae begin to form primordia through their modification, which characterises the reproductive phase. The deposition of new material in the aggregate of hyphae occurs until the mushroom reaches five mm, when the primordium is morphogenetically complete, and from this stage onwards the mushroom elongates and swells, until the spores are released (MOL, 1989).

The growth and development of lignocellulolytic fungi in the different phases depends on the release of hydrolytic and oxidative enzymes that act to depolymerise the constituents of the substrate, making them amenable to absorption by the hyphae (OKEKE et al. 1994).

The absorption of extracellular materials allows the formation of new cells from the uniform deposition of chitin, formed by the incorporation of N-acetyl-D-glucosamine molecules, with (1-4)-β-glucoside bonds between them, deposited on the lateral walls and septa of the hyphae, in order to organise a rigid cell that resists turgor pressure. The glucosamine molecules are associated with the glucans of the plasma membrane through covalent bonds that determine the degree of rigidity and/or plasticity of the cell (MOL, 1989).

As the development of the vegetative mycelium has different characteristics to the reproductive mycelium, the glycosaminoglucans of the vegetative cells differ in their chemical and structural constitution from those of the fruiting cells, depending on whether or not the structure requires cell expansion, i.e. the component cells of the substrate hyphae have more rigid walls, unlike the fruiting hyphae, which require a certain degree of plasticity (MOL, 1989).

The environmental conditions provided during the different phases of the fungi's life cycle can alter the structural composition and consequently the plasticity of the cells, especially the mushroom-forming ones, because in the presence of high $CO_2$ concentrations, changes occur in the bonds between the p-glucans of the cell wall and the elongation of the stipe is impaired, forming mushrooms with reduced stipes (MOL, 1989).

Since the transition from the vegetative to the reproductive phase depends on changing the environmental conditions necessary for mushroom formation in the vegetative hyphae, under these conditions there is a structural reorientation between glycosaminoglucans and β-glucans. Specifically, the glucans acquire a longitudinal arrangement and the glycosaminoglucans a transverse arrangement, a condition that restricts the diameter of the hyphae, allowing the mushroom to grow vertically in certain regions of the mycelium (MOL, 1989).

During the reproductive phase, fungi mainly use the sugars available in the substrate in the form of cellulose, glucose, among others, to form mushrooms and during the vegetative phase, in many cases, lignin is oxidised preferentially to other polymers (DURRANT et al. 1991).

The amount of biomass formed, both vegetative and reproductive, depends, among other things, on how the substrate is processed and its physical, chemical and microbiological composition.

Vegetative biomass can be monitored "in *aitro*" by measuring radial growth in Petri dishes and linear growth in mycelial tubes (NOBLE et al. 1995; SAVOIE et al. 1996). However, the biomass formed may not be directly related to linear growth in cases such as A. bras/l/ens/s, which forms aerial mycelium in culture medium (NEVES, 2000).

Indirect methods for assessing vegetative biomass are recommended due to the difficulty of separating the mycelium from the substrates (SEITZ, 1979). These techniques can use structural or metabolic chemical components of the fungi, such as the quantification of glucosamine from the acetylation of chitin, DNA, RNA and other proteins. In many cases, proteins with extracellular activity are used, such as the laccase enzyme. Oxygen consumption or the release of carbon dioxide, as well as heat production, resulting from metabolic activity are also used (KEREM et al. 1992; MATCHAM et al. 1985; MUDGETT, 1986).

Techniques with greater reproducibility and sensitivity, such as HPLC (high performance liquid chromatography), have been used to monitor the vegetative growth of various groups of fungi (BARAJAS-ACEVES et al. 2002). This method of assessing vegetative biomass in fungi is possible due to the specificity of the marker used, which makes it possible to eliminate interference from other substances. In the case of processed lignocellulosic substrates, traditionally used for Agar/cus cultivation, there are microorganisms that can interfere with the results. To prevent this from happening, an evaluation of the substrate in the absence of mycelial growth should be carried out in order to eliminate the effect of the microbiota present (NEVES, 2000).Ergosterol is the predominant sterol in fungi and the main component of membranes, and has been used to monitor the vegetative growth of various fungi on solid substrates (NYLUND & WALLANDER, 1992; WANG et al. 1995). This sterol is the most sensitive marker, even with small amounts of biomass, and can be detected after seven days of incubation (MATCHAM et al. 1985). In A. blazei (Murrill), ergosterol (Figure 1) was predominant after two weeks of mycelial growth (HIROTANI et al. 2002).

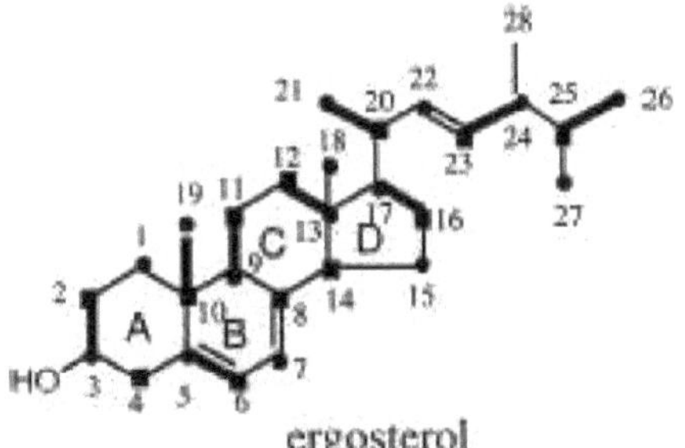


Figure 1. Structural configuration of the ergosterol molecule (HIROTANI et al. 2002).

Neves (2000) during the vegetative growth of A. brasiliensis found values between 163.62 and 187.80 i g of ergosterol per gram of dry composted substrate, similar to the values found by Macauley and Bethel (1990) for A. brunnescens.

**3.3** Growing **substrates**      - selection and treatments

The use of lignocellulosic waste as a substrate for mushroom cultivation requires it to be available in the region in order to reduce production costs and make it easier to repeat the formulation between productions. It is important to emphasise that the formulation, depending on the bulk material used, directly affects productivity (EIRA, 2003). The physical-chemical and degradative characteristics of the materials, as well as their adaptability to the production process, must also be taken into account.

The compositional and degradative characteristics of these materials, as well as the substrate

7

preparation processes, can interfere with the fungus' growth rate. Composted substrates are traditionally used to grow Agaricus, but other alternatives can be used. As a result, the availability of nutrients, especially carbon sources, can be diverse and thus promote different growth rates, both vegetative and reproductive.

In the case of composted substrates, nitrogen, which is necessary for protein synthesis among other things, comes from the reaction between humic substances and nitrogenous compounds during composting, resulting in sources available for absorption. These reactions are partly responsible for the selectivity of the compost, which is not the case when preparing substrates for axenic cultivation (SCHISLER, 1982).

Secondary decomposer fungi, such as Agaricus, are cultivated on previously degraded materials, where the readily assimilable components, such as carbohydrates and sugars, have been consumed.

by most of the mesophilic microorganisms present in the initial phase of composting. At the end of the mesophilic phase, it is replaced by thermo-tolerant microorganisms that fix nitrogen and promote the selectivity of the compost, leaving recalcitrant constituents and complexed forms of carbon that are only accessible to organisms capable of producing specific enzymes.

In this context, composting is important for the production of Agaricus, as in addition to providing a more selective substrate, it discourages the growth of contaminating and/or competing organisms (SCHISLER,1982). Compost for growing Agaricus generally has a bulky lignocellulosic component based on straw, grass or other fibrous materials, usually with a high carbon content (C) and lower nitrogen (N) and phosphorus (P) content, as well as concentrated components (usually bran and cake), which are incorporated in adequate quantities to achieve C:N:P ratios of 30 to 37:1:0.2, 30:1 in the case of A. brunnescens and 37:1 in the case of A. brunnescens. brunnescens and 37:1 for A. brasiliensis (EIRA, 2003).

Kopytowski Filho (2002) concluded that compost to obtain the highest productivity from A. brasiliensis should have a C/N ratio of around 37:1, contrary to the ratio used by most compost producers, who use the ratio recommended for A. brunnescens, which is narrower (around 17:1).

The micronutrients (mg.kg$^{-1}$ ) K, S, Ca and some trace elements such as Mg, Mn, Zn, Bo, Co, Mo and others are already present in the straw, manure and bran in sufficient quantities to support the metabolism of the microbiota involved in the composting process in phases I and II and during the colonisation of the compost by the fungus (EIRA, 2003).

In general, substrates for fungal growth in a solid state require pre-treatment in order to make the chemical constituents more accessible and their physical structure more susceptible to colonisation by the mycelium (MUDGETT, 1986). This pre-treatment usually involves "fermentation" at room temperature, caused by hydration of the bulky material in order to promote microbial activity.

Composting for mushroom cultivation purposes is a biodegradation process involving native microbiota, accelerated by forced oxygenation and resulting in the humification of lignocellulosic sources and caramelisation of carbohydrates. During this process, there is an increase in the activity of bacteria, mainly actinomycetes, as well as fungi, which basically require oxygen and humidity to utilise the sugars, transforming the growth environment. This accelerated growth results in an increase in temperature due to the activity of the microorganisms, which utilise sugars and readily assimilable substances in the compost. At the end of the process, a selective environment is generated for the fungus of interest, with humidity levels suitable for the growth of filamentous organisms, which synthesise a variety of extracellular enzymes (MUDGETT, 1986; SCHISLER, 1982). The intense microbial activity implies a reduction in the C/N ratio, due to an increase in the protein level and carbon consumption, changes in the pH of the medium, among other factors illustrated in figure 2 (SAVOIE et al. 1996; SCHISLER, 1982).

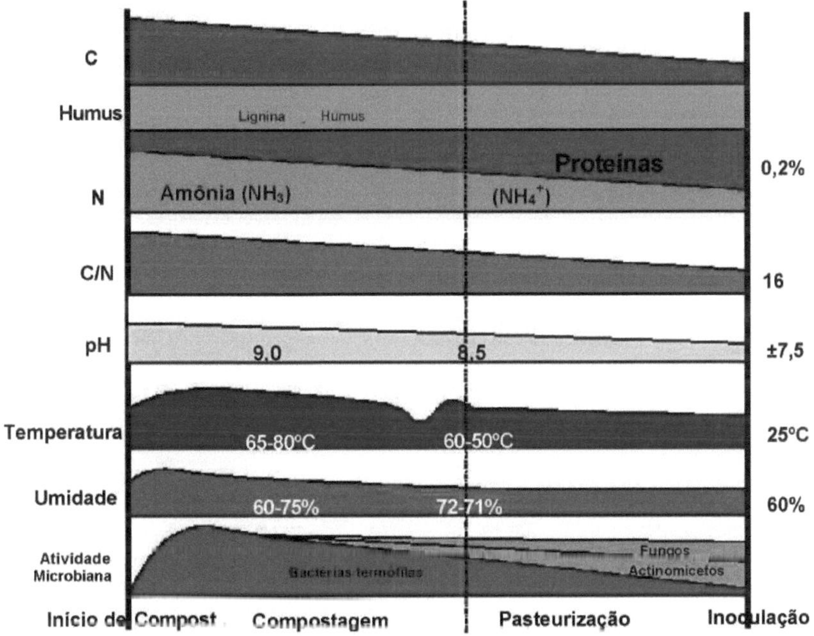


Figure 2: Changes in various factors during the composting process (FERREIR A, 1998).

Regardless of the characteristics of the materials used for the formulation and their proportions in the mixture, the preparation of a compost involves two stages and can take varying amounts of time:

In stage I, also known as Phase I, the ingredients are stacked in layers until they reach a maximum size of 1.80 metres x 1.80 metres, with the length varying. The materials are moistened and supplements are added. The temperature, which varies between 45°C and 80°C in the different parts of the compost heap, also known as the meda, must be monitored. The pile must be turned over periodically to promote a homogeneous environment.

uniform biodegradation of the ingredients, as well as reducing the possibility of anaerobic environments forming (BRAGA et al. 1998).

Temperatures around 80°C contribute to the caramelisation of carbohydrates, one of the processes responsible for the selectivity of compost and the characteristic darkening of the substrate. This process reduces the water content of the carbohydrates, concentrating carbon for later utilisation by the mycelium (SCHISLER, 1982).

The shortening of Phase I is stimulated by the frequency of turning, which promotes aeration, fibre fractionation and accelerates microbial activity during decomposition. Humidity must be strictly controlled so as not to stop microbial activity or promote anaerobic environments. The sooner uniformity is achieved in the substrate, the shorter phase I will be, which can vary depending on the formulation and the characteristics of the materials used. At the end of this phase, the compost has chemical-physical characteristics close to the ideal (BRAGA et al. 1998).

In Phase II, heat treatment consisting of pasteurisation and conditioning, the substrate is transported to closed facilities called tunnels, or pasteurisation chambers, with the aim of eradicating insect eggs and/or larvae, nematodes, among others, brought in from the compost yard, promoting temperatures of 60-64°C / 4-6 hours (Pasteurisation) and between 45-50°C for approximately 7 days (Conditioning) homogeneously throughout the compost mass. This process will eliminate competing fungi and promote the growth of thermophilic organisms that are beneficial to the growth, especially in the vegetative phase, of the fungus of interest (EIRA, 2003; STRAATSMA et al. 1993).

At the end of these processes, the compost for A. brasiliensis should have a humidity of around 65%, a pH of between 6.5-7.5, a C/N ratio of 37/1, total nitrogen of around 1.8 to 2% and the straws should be less rigid with a blackish colour, white-grey hues should be more evident and uniform, indicating correct pasteurisation and adequate conditioning (BRAGA et al. 1998).

During Phase II, nitrogen is fixed by bacteria such as Proteus, Micrococcus and Aerobacter and actinomycetes, thermo-tolerant organisms that transform the volatile nitrogen (NH3 - ammonia) generated by the breakdown of proteins into the ammoniacal form ($NH_4^+$ - ammonium). This process is known as ammonification and leads to a reduction in the ratio between C and N due to the simultaneous consumption of these elements. Thus, when adequate concentrations of carbon and nitrogen are present in the initial composition of the substrate, the release of ammonia occurs in parallel with the decomposition of carbohydrates, otherwise there may be residues of these elements in the final substrate and thus jeopardise mycelial growth and consequently productivity (SCHISLER, 1982).

Volatile nitrogen ($NH_3$) must be completely eliminated during the compost conditioning phase, when the temperature is kept at around 45°C for 6 to 8 days, as volatilisation is favoured from 35°C onwards. This volatilisation is essential because if volatile N is found at levels above 5ppm, it will inhibit the mycelial growth of the fungus of interest (GERRITS, 1988). However, the packaging and transport of the compost after it has been removed from the pasteurisation tunnel must be strictly monitored, as there is a possibility that the forms of nitrogen ($NH_4^+$ to NH3 as per

illustrated below), damaging vegetative growth and mushroom production as well as enabling the development of other competing fungi, such as Coprinus comatus.

The equation for the transformation of nitrogen during composting is as follows:

$$NH3+H2O \; NH4 +^{+-} OH$$
(volatile N)(ammoniacal N)
Ammonia Ammonium

Gerrits (1988) reports that the levels of $NH_4^+$ (ammoniacal nitrogen) directly affect the productivity of Agaricus brunnescens and are directly related to the percentage of total nitrogen at the start of composting (Phase I). Higher yields were obtained with $NH_4^+$ of around 0.4% at the end of phase II, and for these levels to be reached, total nitrogen had to be around 1.5% for classic compost and 2.0% for synthetic compost at the start of phase I.

No bibliographical references were found on the sensitivity of A. brasiliensis to $NH_3$ or on the appropriate levels of $NH_4^+$ for satisfactory yields. However, like champignons, A. brasiliensis is a secondary decomposer that is able to degrade composted substrates rich in nitrogen (STAMETS, 2000). On the other hand, higher yields were obtained in substrates that had been sterilised and not composted, an axenic cultivation system, suggesting versatility of growth in substrates without prior decomposition, indicating possibilities for production in new cultivation technologies (SÀNCHEZ VÁSQUEZ, 1999).

The German researcher Till (1962) was the forerunner of alternative systems for the production of mushrooms of the genus Agaricus, which disconnect the producer from the risks of composting and pasteurisation, through the development of production systems in sterilised substrate. Basically, mushroom cultivation can generally be carried out under axenic conditions, i.e. the substrate must be sterilised and the cultivation techniques are aseptic until the sub-tract is completely colonised (EIRA, 2000).

Mushroom cultivation on sterilised, non-composted substrates was first developed at the Max Planck Institute in Hamburg, Germany in 1970. The main aim of this research was to reduce the uncertainties and risks of the composting process in mushroom cultivation by providing a nutritious substrate that at the same time prevented the action of competing microorganisms, since thermophiles had been inoculated in the quest to achieve selectivity of the growth substrate using alternative technologies (MOLENA, 1986).

Production in sterilised and non-composted substrates, as an alternative to traditional systems, requires greater initial investment, but in the medium term, the technology becomes more economically rewarding, as it requires less labour, the producer verticalises his production and can increase productivity due to the possibility of supplementing the substrate. In addition to these advantages, this production technique can be used to cultivate other species of fungi (SÀNCHEZ- VÁSQUEZ,1999).

Eira (2003) reports that in axenic systems, there is the potential to use a C/N ratio of between 15 and 25/1 in order to obtain greater productivity, which reduces the high costs resulting from the sterilisation process. Sánches-Vàsquez (1999) obtained 48.64% biological efficiency (EB=fresh weight of mushrooms/dry weight of substrate) for A. blazei Murrill in sterilised substrate, with a C/N ratio of 37/1, similar to that obtained in the traditional system where compost is used.

However, axenic systems do not have the same selectivity as that provided by the composting process, because in substrates for axenic cultivation, there are readily assimilated nutrients available in the substrate, which can favour the growth of competing organisms.

In order to minimise risks, the workforce to be used in axenic systems must be specialised, since handling must be carried out under axenic conditions. These procedures require specialised knowledge, especially in the field of microbiology, which further complicates the process.

**3.4Alternatives** for mushroom cultivation in Santa Catarina

The substrates and formulations used to grow mushrooms vary depending on where the crop is grown. Santa Catarina has potential for mushroom production, as it has a diversified agricultural activity, whose residues can be applied to both traditional and axenic cultivation, as well as suitable climatic conditions, enabling the diversification of cultivated species and investment options, which are stimulated through inter-institutional partnerships, which provide technical support to producers and promote the activity in the state.

The utilisation of palm heart production waste from the exotic species popularly known as the Australian royal palm (Archontophoenix alexandrae Mart.) is a promising option, since more than 60% of the palm heart produced in Santa Catarina is from this species and 80% of the harvested plant is discarded, making it an abundant waste product (http://www.an.com.br).

Interest in the production of royal palm arose because of the rapid growth rate of this plant compared to the native species (Euterpe edulis Mart.), and also with a view to reducing extractivism in the state. While the royal palm takes two to three years to produce a kilo of palm heart, the native palm requires seven to eight years to be suitable for cutting (http://www.an.com.br).

The royal palm (A. alexandrae) has been widely cultivated in the tropics and subtropics as an ornamental plant. In the state, plantations of the Australian royal palm can be found in Florianópolis, Sombrio, Praia Grande, Camboriú and the Vale do Itajaí region.

The royal palm leaves removed when the palm heart was cut and harvested proved to be potentially usable as a substrate for mushroom production in an axenic system, mainly because they provided the substrate with a good structure, as well as suitable physicochemical characteristics, since preliminary experiments using this material during the cultivation of A. brasiliensis in an axenic system, 48% biological efficiency was obtained (data not shown) and the median leaf sheaths proved suitable for growing shiitake (Lentinula edodes (Beck.) Pegler) in the same system (TONINI, 2004).

Banana tree waste also has chemical-physical and structural characteristics suitable for use in mushroom cultivation. According to Sturion (1994) this waste represents a proportion equivalent to 40% of the weight of the fruit and the leaves 15% of the whole plant, with an estimated amount of banana waste in Santa Catarina of around 200,000 tonnes/year.

Rice straw is also significant in the state of Santa Catarina, mainly due to its planted area of around 113,256 hectares (AGENDA 21, 1997) and can provide farmers with income in the months between harvests, since after harvesting the grains, its composition is 40-50% cellulose and hemicellulose, 10-20% lignin and 0.5-0.7% nitrogen (KAKEZAWA et al. 1990). As well as being one of the most widely used residues in the world for the production of Agaricus, it can also be used to grow various species of the Pleurotus genus.

Another material available in the state of Santa Catarina is sawdust, mainly from Eucalyptus species, which is currently used as an integral part of substrates for growing Lentinula edodes and Pleurotus species (CAMPBELL & SLEE, 1987), as well as for Agaricus in an axenic system (SÀNCHEZ-VÁSQUEZ, 1999). Its availability is due to the fact that it represents 8% of the material processed (800,000m$^3$ of logs/year in Santa Catarina) and the existence of approximately 4,000 timber industries in the state, mainly in the mountainous region.

Agaricus producers, however, do not consider the issue of regional availability of waste by using formulations, substrates and treatments typical of temperate regions. This is due to the scarcity of technical and scientific information on the possibility of using new formulas, as well as alternative production techniques for mushroom cultivation, making it feasible to use local waste.

3.    5Degradative capacity of basidiomycetes

White rot fungi, known as "White-rot", degrade all plant cell wall structures, with the speed with which each of these constituents is degraded differing, unlike brown rot fungi or "Brown-rot", which selectively degrade carbohydrates, with limited degradation of lignin, which remains in large quantities in the substrates colonised by these fungi (PANDEY & PITMAN, 2003).

Organisms belonging to the basidiomycete class, such as A. brasiliensis, generally have an enzyme complex made up of cellulases, ligninases, peroxidases, proteases, phenol oxidases and other enzymes capable of degrading recalcitrant components and utilising the carbon sources of these constituents (BUSWELL et al. 1996).

Both cellulose and lignin have carbon as the basis of their molecular structures, making them important in the cycle of this element. Lignin, together with hemicellulose, surrounds cellulose fibres, becoming a physical barrier to the degradation of cellulose. The solubilisation of lignin may occur primarily to allow access to carbohydrates, since it does not serve as an energy source and requires a substantial energy investment, due to the fact that lignin is associated with polysaccharides and is probably covalently linked to hemicellulose (BUSWELL et al. 1996).

Due to the structural variability of these elements, the speed at which they degrade in nature also varies. Hemicellulose, the intermediate layer between lignin and cellulose, is degraded in a shorter time, around two years. Cellulose, on the other hand, is degraded over a longer period of time, around three and a half years. However, lignin is the component with the longest degradation time.

This is due to its three-dimensional structure, where there are no repeated bonds in the construction of this phenylpropane heteropolysaccharide (BUSWELL et al. 1993, 1996).

Lignin is a recalcitrant constituent that limits the attack sites of cellulose, and despite its high calorific potential, there is no known organism capable of using it as its sole source of energy, which results in its accumulation in nature (RAJARATHNAM et al. 1992 apud SANTOS, 2000).

In this way, the utilisation of the polysaccharides present in lignocellulosic waste is limited by the presence of high amounts of lignin (KEREM & HADAR, 1993). Among the oxidative enzymes involved in the degradation of phenolic and non-phenolic lignin are laccase (Lac), lignin peroxidase (LiP) and manganese peroxidase (MnP). Unlike hydrolytic enzymes, which act by separating cellulose into glucose monomers, oxidative enzymes act by forming radicals that will destabilise the lignin molecule (BUSWELL et al. 1993, 1996).

Oxidising enzymes act by removing the methoxyl groups from the lignin and breaking the ether bonds, releasing more soluble substances and exposing the remaining chains to the action of other enzymes (BUSWELL et al. 1996).

The production of these enzymes by the fungal mycelium is a crucial activity during the colonisation process and an important determinant of mushroom yield (BUSWELL et al. 1996).

Basidiomycetes are the fungi most capable of degrading lignin and are therefore able to use the carbon in lignocellulose. However, there is a certain energy demand for this use, which may be responsible for the activation of secondary metabolism in these fungi, i.e. the production of oxidative enzymes due to an unfavourable environment for growth or limited carbon sources (KEREM et al. 1992).

Laccase can participate in the degradation of lignin by oxidising phenolic groups, however, this may imply a limited role for this enzyme in this function, since the phenolic subunits make up a small portion of this polymer (BONNEN et al. 1994).

The role of laccase in lignin degradation is still poorly understood. Lacase is credited with detoxifying some low molecular weight phenolic monomers that originate from lignin degradation and can inhibit fungal growth (CAI et al. 1998). Therefore, fungi that produce oxidative enzymes such as lacase have environmental advantages over strictly cellulolytic fungi such as Volvariella volvacea (BOLLAG et al. 1988 apud CAI et al. 1998).

Manganese peroxidase (MnP), in turn, acts by mediating the initial stages of degradation of high molecular weight lignin, transforming Mn (II) into Mn (III), which can act directly in the oxidation of phenolic groups or in other compounds, since it is non-specific (HATAKKA, 1994).

Since the 1980s, when MnP and LiP peroxidases were discovered, certain groups of fungi, especially decomposers, could be classified according to their enzyme production: those that produce LiP - MnP, MnP - Lac and LiP - Lac, although some exceptions do occur (HATAKKA, 1994).

Several fungi that are efficient at degrading lignin in nature apparently do not produce LiP, but this enzyme has been shown to be responsible for oxidising phenolic lignin (HATAKKA, 1994). A. brunnescens is a lignin-degrading fungus, but no LiP activity was detected when veratryl alcohol was used as the enzyme substrate, the standard method for assessing the activity of this enzyme (BONNEN et al. 1994).

3.2.1 Utilisation of polysaccharides

The efficient utilisation of cellulose by degrading fungi requires the cooperative action of several hydrolytic enzymes, which hydrolyse polysaccharides into monomers that can be absorbed. The enzymes responsible for this process are cellulases, endo- and exoglucanases, |3-glucosidase and cellobiohydrolase, which degrades cellobiose, one of the products of cellulose hydrolysis and an inhibitor of cellulase (CAI et al. 1998, 1999; DING et al. 2001).

Cellulose is the most abundant organic compound in nature and the predominant polysaccharide in plant waste, accounting for between 30 and 60 per cent of its total dry weight, and should therefore be used more efficiently in view of the increase in food demand resulting from the world's growing population (REID, 1989).

Decomposing fungi efficiently use cellulose as a carbon source, mobilising it mainly for the production of fruiting bodies, as in the case of A. brunnescens (WOOD & GOODENOUGH, 1977) and Pleurotus spp. (VELÁZQUEZ-CEDENO et al. 2002).

Chemical analyses in compost for growing A. brunnescens after applying the substrate cover layer to induce fruiting showed preferential degradation of cellulose and hemicellulose for fruiting, with an increase in endoglucanase production and a reduction in lignolytic activity.

However, lignin was preferentially degraded before covering the substrate (DURRANT et al. 1990; WOOD & GOODENOUGH, 1977).

Hemicellulose accounts for up to 40% of the dry weight of lignocellulosic waste and its degradation is carried out by hemicellulases such as xylanase, which hydrolyse this heteropolysaccharide with short branched chains made up of hexoses, pentoses, uronic acid and smaller sugars.

The removal of carbon through xylanase activity in A. brunnescens was not related to its mobilisation for fruiting (WOOD & GOODENOUGH, 1977), and could therefore be a source of carbon for vegetative growth.

Other hydrolytic enzymes such as β-glucosidase have been extensively studied for their potential to convert cellulose into glucose on an industrial scale, as well as for their ability to act on a wide range of carbon sources (CAI et al. 1998), although their role in mobilising carbon for fruiting is unknown and as far as the genus is concerned has been little studied.

Bearing in mind that mushroom production and the conversion rate depend primarily on the efficient colonisation of the substrate and its degradation by the mycelium through the fungal enzyme pool (SAVOIE et al. 1996), it is assumed that enzyme expression and activity vary, above all depending on the species of fungus used, the availability of cellulosic material in the substrate, the presence in greater or lesser quantities of recalcitrant constituents, as well as the heat treatment of the substrate and the environmental growth conditions, factors that directly influence fruiting yields (VELÁZQUEZ- CEDENO et al. 2002).

3.5.2  Substrate degradation dynamics studies

One of the most important factors in the degradation of the components of a lignocellulosic substrate is the production of extracellular enzymes by the mycelium to nourish the fungus during the vegetative and reproductive phases (VELÁZQUEZ-CEDENO et al. 2002) and the variations in enzymatic activity related to the presence of greater or lesser amounts of inducing or inhibiting substances in the substrate (BUSWELL et al. 1993).

If the substrates used for cultivation are nutritionally unbalanced, they promote metabolic alterations in the fungi and can reduce lignocellulose degradation. Excess carbon in the medium represses the degradation of lignin by oxidative enzymes, as they result from the fungi's secondary metabolism. On the other hand, high concentrations of nitrogen do not reduce the rate of lignin degradation in Pleurotus ostreatus and Lentinula edodes (HATAKKA, 1994), but can impede the growth of the fungus of interest due to the appearance of contaminating competing organisms.

Therefore, in addition to favouring productivity, substrates with balanced formulations allow the fungus to act in such a way as to obtain the necessary nutrients at the right time, i.e. the degradation of lignocellulose occurs according to the incubation time, i.e. at each stage of development in the crop cycle.

In the case of Pleurotus sp. the decrease in laccase and Mn Peroxidase activity is concomitant with the development of fruiting and the increase in cellulase activity. In addition, the relationship between laccase and cellulase activity is inverted after harvest and returns to this condition with the appearance of new fruit (VELÁZQUEZ-CEDENO et al. 2002).

In the cultivation of A. brunnescens, laccase activity is high during the formation of the primordia, during which time there is a reduction in lignin content. However, its activity decreases rapidly with the formation of the aggregate of hyphae, which will differentiate to form the mushroom primordia. The peak in cellulase activity occurs during the development of the fruiting bodies, a situation that is characteristic of various species of cultivated fungi (KUES & LIU, 2000; BONNEN et al. 1994).

As well as defining the dynamics of the enzymatic activity responsible for the changes in the substrate, the study of the changes themselves, using different methodologies, provides complementary information on the real consumption of nutrients and the degradative activity.

Different methods of analysis have been used to monitor lignocellulosic degradation by fungi. Traditional methods such as chemical dosages are being complemented and/or replaced by less costly procedures that use smaller quantities of sample and produce results in a shorter space of time (PANDEY & PITMAN, 2003; YANG et al. 1993). Chemical methodologies such as the determination of lignin using the

Klason methodology are common for assessing the concentration of lignin present in the substrate (TUOMELA, 2000).

Yang et al. (1993) verified the utilisation of lignocellulose by Pholiota nameko using a chemical methodology, isolating lignin and dosing carbohydrates, making a relationship between the reduction in substrate weight associated with reproductive biomass and consequent use in respiration. This study found that the degradation of lignin and hemicellulose was faster than cellulose during mycelial growth at the end of 199 days, but during the last 30 days, between 200 and 230 days, cellulose was consumed more quickly. This peak in cellulase production corresponded to the period of formation and development of the fruiting bodies. At this point, lignin and hemicellulose had already been reduced by around 80 per cent.

Alternative methodologies to traditional chemical methods can be used to assess the degradation of lignocellulosic substrates. Analyses using infrared spectroscopy are routinely carried out to study lignin (FAIX, 1986), using a spectrometer operating in Fourier transform mode, known as FTIR (CHEN, 2000).

Infrared spectroscopy provides information on the presence or absence of functional groups in various organic and inorganic materials, as well as their chemical structure. Changes in the relative intensity of the spectral bands indicate changes in the chemical structure of the substrate (www. micromemanalytical.com).

Using FTIR, Chen et al. (2000) observed changes in compost during the growth of Agaricus brunnescens, finding a similarity between the spectra of the inoculation period and substrate coverage. However, the greatest changes occurred during fruiting, when there was a reduction in polysaccharide peaks (1000 - 1100 $cm^{-1}$), an increase in the levels of aromatic structures resulting from oxidation processes (1650 $cm^{-1}$) and COO ions$^{-}$ (1325 $cm^{-1}$), probably linked to the oxidation of lignin side chains, a classic behaviour of lignocellulose degradation.

Several studies have described qualitative changes in FTIR spectra in materials exposed to degradation for long periods of time, however, information on relative changes in the lignin/carbohydrate composition of wood, as well as other materials exposed to short periods of degradation, is scarce (PANDEY & PITMAN, 2003).

# CHAPTER 2

**4.** Lignocellulolytic enzyme activity profile and biomass production during vegetative growth of Agaricus brasiliensis in two cultivation systems: traditional and axenic

## Summary

The use of lignocellulosic substrates by certain groups of fungi depends on their ability to produce hydrolytic and oxidative enzymes, release them into the extracellular environment and utilise the nutrients that are essential for vegetative and reproductive growth. Thus, the enzyme profile, the dynamics of enzyme activity and the corresponding production of vegetative biomass can provide important information on the efficient utilisation of these substrates. The profile was defined and the enzymatic activity of Agaricus brasiliensis in traditional and axenic cultivation was evaluated, as well as the production of mycelial mass. A. brasiliensis produced various enzymes such as laccase, manganese peroxidase, B-glucosidase and xylanase. No lignin peroxidase or cellulase (FPA) activity was detected. Lacase activity was higher when compared to Mn peroxidase, especially in axenic cultivation. The enzyme xylanase showed higher levels of activity in composted substrate. The formation of greater mycelial mass in traditional cultivation than in axenic cultivation indicates that the fungus grows faster in degraded substrates, with significant differences in biomass formation in both substrates from the 16th day of growth onwards. Laccase activity and protein biosynthesis were correlated with vegetative biomass in composted substrates. A. brasiliensis can grow on both types of substrate, with differences in the utilisation of lignocellulosic components in relation to degradation and bioconversion.

Key words: ergosterol, Agaricus brasiliensis, enzymes, lacase, vegetative growth

## 4. 1INTRODUCTION

Most commercially cultivated basidiomycetes convert lignocellulosic waste into mushrooms that can be used for human consumption, while the residual substrate can be used as animal feed and for environmental bioremediation (LAW et al. 2003; VILLAS- BÔAS et al. 2002).

Mushroom cultivation is one of the most efficient strategies for utilising agro-industrial waste and transforming it into high value-added nutraceutical and nutritional products. However, the efficient use of a lignocellulosic substrate is directly related to the ability of these organisms to metabolise the material and produce fruiting bodies.

The use of lignocellulosic substrates as a carbon source for conversion into fruiting depends on the ability of the fungus to produce hydrolytic and oxidative enzymes and release them into the extracellular medium, as well as producing competitive vegetative biomass, since the success of colonisation of a substrate can be determined by competitive ability and inoculum potential in relation to the inoculum potential of competitors (BUSWELL et al. 1996; SAVOIE et al. 1996).

Rapid vegetative growth reduces the incidence of disease in the culture, making the fungus better able to compete and allocate resources through appropriate enzymes, especially in compost, where pasteurisation only partially reduces the total microbiota. This compost production strategy is currently used in the cultivation of Agaricus brunnescens Peck [= Agaricus bisporus (Lange) Imbach] and also in the cultivation of A. brasiliensis S. Wasser et al. (2002) = (A. blazei Murrill ss. Heinemann).

Both vegetative and reproductive growth depend primarily on the degradation of the main components of lignocellulosic material, i.e. lignin, cellulose and hemicellulose. However, the utilisation of the carbon sources of these materials (cellulose and hemicellulose) depends on the degradation and/or modification of the external recalcitrant material, lignin.

Some basidiomycetes such as Agaricus have oxidative enzymes that act in the biodegradation or biotransformation of lignin, however, the regulation of enzyme production as well as the exact role of these enzymes during the fungi's cultivation cycle is still unclear. Fungi such as Ceriporiopsis subvermispora produce Lacase and MnP (manganese peroxidase) and have genes for the production of LiP (lignin peroxidase), however, the activity of this enzyme has not yet been detected (RUTTIMANN et al. 1993).

Based on the types of oxidative enzymes produced, fungi can be classified into three groups: a) LiP and MnP producers such as Phanerochacte chrysosporium, b) MnP and Lac (laccase) producers such as A. brunnescens and c) LiP and Lac producers such as Phlebia ochraceofulva, however, exceptions do occur. Recently, a new group represented by Pleurotus spp. has been identified as laccase and AAO (aryl alcohol oxidase) producers (HATAKKA, 1994).

In an experiment using synthetic lignin as a substrate, Hatakka (1994) found that fungi belonging to group a, which produce LiP and MnP, were the most efficient at degradation. On the other hand, fungi that produce Lacase and MnP, as in the case of Agaricus , showed moderate to good levels of synthetic lignin degradation using these enzymes.

In the case of Rigidiporus lignosus, these enzymes act synergistically during lignin degradation (GALLIANO et al. 1991; HATAKKA, 1994).

High levels of laccase activity followed by MnP were observed during the vegetative growth of A. brunnescens in commercial compost, with lignin being preferentially degraded during this phase (DURRANT et al. 1991). However, when substrate mulch was added to induce fruiting, changes were observed in the level of activity of these enzymes, with a reduction in lignolytic activity followed by a subsequent increase in cellulolytic activity (WOOD & GOODENOUGH, 1977).

Laccase is the extracellular protein predominantly produced by A. brunnescens, and has been used as an indirect biomass marker, since when low levels of activity are detected, a low amount of biomass is produced (BONNEN et al. 1994; MATCHAM et al. 1985; WOOD & GOODENOUGH, 1977).

The amount of total biomass produced reflects the fungus's ability to utilise carbon sources, as well as the energy expenditure to access it. To monitor the vegetative growth of fungi that grow on solid substrates, other specific markers can be used, such as structural elements or indirect markers like extracellular proteins (MATCHAM et al. 1985). Among the constituent markers, ergosterol is considered the most sensitive, since it is the predominant sterol after two weeks of A. blazei cultivation (HIROTANI et al. 2002).

Enzymatic activity and biomass production can vary depending on the growth substrate and also on the treatment this material received before the fungus grew.

Composted lignocellulosic substrates are pre-degraded and undergo a series of transformation reactions that affect the shape and availability of nutritional resources (GERRITS, 1988). On the other hand, axenic substrates are not subjected to prior degradation treatments, so the lignocellulose maintains its structure intact, which can lead to changes in the dynamics of the fungus's enzyme production under these conditions.

Depending on their enzyme profile, a relationship can be established with the fungus' ability to grow on these substrates. For example, fungi that don't produce oxidative enzymes may have environmental disadvantages compared to those that do, because the increase in vegetative biomass indicates an increase in the degradation rate and consequently leads to the production of auto-inhibitory by-products, such as phenols, which are also substrates for oxidases, especially laccase (BOLLAG et al. 1984).

During the life cycle of these fungi, not only are oxidative enzymes fundamental, but hydrolytic enzymes, which are responsible for carbon mobilisation, are also regulated depending on the fungus' stage of development. Strains of A. brunnescens genetically modified not to fruit showed no increase in the production of hydrolytic enzymes over 60 days under cultivation conditions. However, in the unmodified strains, the increase in cellulase and endoglucanase activity indicated that the carbon sources were mobilised by the action of these enzymes. Therefore, hydrolases are necessary for mobilising the carbon required for the formation of fruiting bodies, which occurs after the substrate coating has been applied. On the other hand, the enzyme xylanase, which acts on hemicellulose, did not correspond to fruiting and may be responsible for mobilising carbon during vegetative growth (WOOD & GOODENOUGH, 1977).

Enzymatic activity can vary depending on the composition of the growth substrate, as the presence of inducing or inhibitory substances, as well as the stage of initial degradation, can be determining factors in its level of activity. Composted substrates have been used in recent decades to grow Agaricus. However, the preparation of this type of substrate, although widely used, is a complex activity and some of its stages are still not fully understood. In Brazil, commercial composted and pasteurised substrates used for growing Agaricus have different formulations depending on the region, and those produced for growing champignons and used for A. brasiliensis are often of poor quality. In this context, the productivity of A. *brasiliensis,* which often shares the same type of substrate, rarely exceeds 15% (fresh mushroom/wet substrate) in composted substrates with a C/N ratio of 20/1 and total N of 1.8%, unlike *A. brunnescens,* which commonly reaches 25% productivity (fresh mushroom/wet substrate) (FERREIRA, 1998).

When comparing the nitrogen content in the fruiting bodies of *A. brunnescens* and *A. brasiliensis, it can be* seen that this value is markedly different and, therefore, it is important to provide the correct amount of nitrogen required for each species in each substrate, and the preparation of their growing substrates should be disconnected, thus defining specific substrates and formulations capable of promoting higher yields of *A. brasili ensis.*

The difference in the chemical composition of the fruiting of these fungi, the substrates used for production and the environmental conditions required have led to an investigation into the potential of A. brasiliensis to produce enzymes and vegetative biomass in traditional composted substrate and in axenic substrate as an alternative for obtaining fruiting. When grown in non-composted axenic substrate, A. blazei Murrill produced fruit with 43.6% biological efficiency (SÀNCHEZ- VÁSQUEZ, 1999), using a methodology adapted from Till (1962). Non-composted substrates are an alternative for using lignocellulosic waste in an attempt to eliminate the complexity of the composting process, and are also a common production technology for other species, as in the case of Pleurotus spp. and Lentinula edodes . This aspect could make axenic cultivation of A. brasiliensis viable for companies that have the appropriate infrastructure for this and other species.

These were the aims of this work:

- To evaluate the production of some lignocellulosic enzymes and the enzymatic dynamics in the degradation of composted and non-composted substrates;

- To evaluate the production of vegetative biomass produced by A. brasiliensis in composted and non-composted substrate during 35 days of growth;

- Identify possible indirect markers (laccase and total proteins) for monitoring vegetative biomass;

## 4.2  MATERIALS AND METHODS

**Fungus:** The strain used was deposited in the fungi collection of the Edible and Medicinal Mushrooms Laboratory at UFSC, Department of Microbiology and Parasitology, under code UFSC-51 and identified as Agaricus brasiliensis = (Agaricus blazei Murrill ss. Heinemann), after taxonomic, genetic and ecological studies by Neves (2000). This strain has been used in commercial cultivation by various producers in the southern states of Brazil.

**Seed inoculum:** The UFSC-51 strain was grown in Petri dishes in MLA medium (30g Malt extract, 5g Yeast extract and 20g Agar/L) at $26\pm1°C$. After 10 days, % of *each* agar plate colonised by the fungus was transferred to 1000mL screw-top glasses containing 350g of boiled wheat grains sterilised in an autoclave at $121°C$ under a pressure of $1.3Kg/cm^2$ , for 1 hour, and incubated for mycelial growth for 20 days at $21\pm2°C$. At the end of this period, the seed was completely colonised by the fungus and was used to inoculate the substrates.

**Substrate preparation and inoculation:** The composted substrate was produced in a commercial composting plant in São José dos Pinhais, PR - Brazil and its formulation was based on maize straw and cobs, straw from various grasses, chicken manure, urea and gypsum, with a final C/N (carbon/nitrogen ratio) of around 20/1 and an initial pH of 7.3. The non-composted substrate was prepared using the following formulation (w/w): 50% royal palm leaf (Archontophoenix alexandrae Mart.), 25% Eucalyptus sp. sawdust, 12% wheat bran, 8% $CaCO_3$, 5% $CaSO_4$, with a C/N ratio of 25/1 and an initial pH of 6.9 (table 1). The substrates were packed in transparent bags of

polypropylene bags measuring 45x20cm, specifically designed for mushroom cultivation, with filters to allow for air exchange. Twenty bags were prepared for each of the two treatments, containing 2kg of substrate with 67% humidity (assessed by the difference between the initial weight and the final weight at 65°C, until constant weight). Both substrates were sterilised simultaneously in an autoclave at $121°C$, under a pressure of $1.3Kg/cm^2$ , for 1 hour in order to eliminate cellulolytic microorganisms that could interfere with the enzymatic evaluations, particularly in the composted substrate. The substrates were inoculated under aseptic conditions in a laminar flow cabinet with 1.5% (w/w) seed-inoculum. Twenty bags (sample units) per treatment were placed on benches in the grow house. The design was randomised and the bags were incubated at 25±4°C with a relative humidity of 73±5% for 35 days.

Table 1- Formulations of composted and non-composted substrates used for the vegetative growth of A. brasiliensis for 35 days

| Ingredients (A) Compost | Quantity | Ingredients (B) Not composted | Quantities % (Kg) |
|---|---|---|---|
| Straw and corn cobs | 2,000 tonnes. | Palm leaf | 50 (7) |
| Various grasses | 4,000 tonnes. | Sawdust | 25 (3,5) |
| Chicken manure | 2,900 tonnes. | Wheat bran | 12 (1,7) |
| Soya bran | 250kg | CaCo $_3$ | 8 (1,4) |
| Ammonium nitrate | 50kg | Plaster | 5 (0,7) |
| Urea | 50kg | | |
| Plaster | 400kg | | |

**Sampling:** 3 bags of each treatment were drawn every 7 days (triplicate analysis). Four sub-samples were taken lengthways from each cultivation bag (sampling unit), using a tube with a sharp end, measuring 30x3.5 cm, which were then mixed together. The samples were weighed and 3g of each sample unit was used for the enzymatic study immediately after sampling; the rest was dried at 55°C for 72 hours. Of the rest, 1g of each sample unit, previously homogenised, was used for biomass analysis (ergosterol) at times 1, 3 and 6 (days zero, fourteen and thirty-five, respectively).

### 4.2.1 Evaluation of the enzyme profile and activity

To each 3g of fresh substrate per sample unit, 30pL of triton X100 and 30mL of ultrapure water were added. The material was macerated, filtered and centrifuged at 9,000g for 15 minutes at 4°C, and the supernatant was used immediately for the enzyme assays. The procedures for assessing enzymatic activity were as follows for each enzyme:

Manganese peroxidase: the activity of manganese peroxidase (MnP, EC 1.11.1.13) was measured using phenol red as a substrate at 610 qm ( $\hat{A}610 = 4460$ M$^{-1}$ cm$^{-1}$ ), according to the methodology of Kuwahara et al. (1984). One unit of enzyme activity was defined as the amount of enzyme capable of oxidising 1 mmol (millimolar) of substrate per minute. A sample without manganese was made to demonstrate the dependence on this element.

Lignin peroxidase: lignin peroxidase activity (LiP, EC 1.11.1.14) was determined by monitoring the oxidation of veratryl alcohol to veratraldehyde at 37°C, indicated by an increase in absorbance at 310 qm ($\hat{A}$

310 = 9300 $M^{-1}$ $cm^{-1}$ ) according to the methodology of Tien and Kirk (1984). One unit of enzyme activity was defined as the amount of enzyme capable of oxidising 1 mmol of substrate per minute.

Lacase: Lacase activity (EC 1.10.3.2) was monitored at 525 qm ($\hat{A}$ 525 = 65000 $M^{-1}$ $cm^{-1}$ ) through the oxidation of syringaldazine. One unit of

enzyme activity was defined as the amount of enzyme that oxidised 1 mmol of substrate per minute.

Cellulase: The activity of cellulase in paper fibre (FPase) (EC 3.2.1.4) was determined by hydrolysis of Whatman N°1 paper, according to the methodology of Mandels et al. (1976). One unit of enzyme activity was defined as the amount of enzyme capable of catalysing the release of 1 mmol of glucose per minute at 50°C. The standard curve was obtained using glucose in different concentrations.

B-glucosidase: B-glucosidase activity was measured by monitoring the release of p-nitrophenyl from p-nitrophenyl-B-D-glucopyranoside at 410 nm ($\tilde{A}$ 410 = 18.5 $M^{-1}$ cm ).$^{-1}$

Xylanase: xylanase activity (EC 3.2.1.8/EC 3.2.1.32/EC 3.2.1.136) was determined by measuring the amount of reduced sugars released from xylan in 2% acetate buffer, using the method described by Bailey et al. (1992). One unit of enzyme activity was defined as the amount of enzyme that catalyses the release of 1 mmol of xylose per minute at 50°C. The standard curve was obtained using xylose in different concentrations. The total sugars reduced through enzymatic action were determined using the dinitrosalicylic acid (DNS) method (MILLER, 1959).

### 4.2.2 Evaluation of     vegetative biomass (Ergosterol and total proteins)

The ergosterol content was assessed using the method described by Barajas-Aceves et al. (2002) with modifications. The dried substrates were weighed (1.0g) into centrifuge tubes, methanol (10mL) added, mixed in a vortex for 1 min and ultrasonicated for 3 min in ice water. The tubes were kept on ice for 15 minutes and finally centrifuged at 12,000 rpm for 15 minutes at 4°C.

The biomass precipitate was washed three times with 10mL of methanol. The supernatants were transferred to 200mL flasks with screw caps. Then absolute ethanol (5mL) was added along with 2g of KOH as a saponifying agent. The methanol extracts were saponified under reflux in a water bath at 80°C for 1 hour. After cooling, hexane (10mL) was added to the mixture and the flasks were then closed and shaken vigorously.

Two phases were obtained and separated in a separating funnel. The alcoholic phase was washed twice with 10mL of hexane and discarded. The two hexane phases from each flask were combined and evaporated overnight. The extracted ergosterol was redissolved in 1mL of methanol for HPLC (high performance liquid chromatography) and immediately used for injection into the system, in a quantity of 100l L of suspension. The extracts were analysed on a Shimadzu apparatus (LC-10AS) equipped with a reverse phase column (C18 Supersil ODS 150 x 4.4 mm). The samples were eluted with a mobile phase of methanol-water (95:5) at a flow rate of 0.6mL min$^{-1}$ , monitored at 282 çm on a UV-VIS detector (Shimadzu SPD-10A model).

The efficiency of ergosterol extraction was determined by adding a known quantity of recrystallised ergosterol (Sigma) to non-colonised substrate, just as for the standard curve (Appendix A).

The content of total soluble proteins in the extracts was determined according to the colourimetric method described by Bradford (1953), with bovine serum albumin (BSA) used as the standard.

## 4.3 RESULTS AND DISCUSSION

### 4.3.1. Enzyme profile and dynamics in traditional and axenic cultivation of A. brasiliensis

There was no difference in the enzyme profile of Agaricus brasiliensis grown on composted and non-composted substrates, producing the following enzymes: laccase (Lac), manganese peroxidase (MnP) (oxidative), xylanase (Xil) and â-glucosidase (hydrolytic). The activity of cellase-FPA (FPA-Cel) and lignin peroxidase (LiP) was not detected. However, Bonnen et al. (1994) indicated that in the case of Phanerochaete chrysosporium the non-detection of LiP activity was due to the presence of an inhibitory factor in the extract. Using immunological techniques, these authors verified the presence of the LiP isoenzyme, a fact that suggests the need to use other methods, such as immunological and/or genetic ones, to verify the fungus's potential to produce this enzyme.

According to Hatakka (1994), Agaricus brasiliensis can be included in group B depending on the type of lignolytic enzymes produced, and the author indicates that in the case of white rot fungi, only two of the three known oxidative enzymes are generally produced. The presence of laccases and either isolated or associated with another peroxidase has been demonstrated in a variety of white rot fungi (MAYER & STAPLES, 2002).

The fact that LiP activity was not detected does not imply that lignin degradation does not occur, as Durrant et al. (1991) showed that in the cultivation of A. brunnescens no activity of this enzyme was detected, however, there was degradation of this polymer, especially during the vegetative phase, when the production of the extracellular protein lacase predominated. Although the role of this enzyme in lignin degradation is not clearly defined, the lacases produced by P. chrysosporium and Coriolus versicolor are involved in the polymer degradation process (ANDER & ERIKSSON, 1976; KAWAI et al. 1988 apud DURRANT et al. 1991) and, in other fungal species, are also related to the regulation of morphology, virulence control and nutrition (MAYER & STAPLES, 2002).


Laccase activity (Figure 3) was predominant among the oxidative enzymes in the vegetative phase, as occurs with A. brunnescens, which is estimated to produce 2% of total proteins in the form of this enzyme (WOOD & GOODENOUGH, 1980). After the seventh day of inoculation, this enzyme showed greater activity in the non-composted substrate, reaching a maximum activity value (138.66 U = enzyme activity units) at the end of 14 days, remaining high (132.21 U) until 21 days of incubation. After this period, enzyme activity gradually decreased to 52.42 activity units (U).

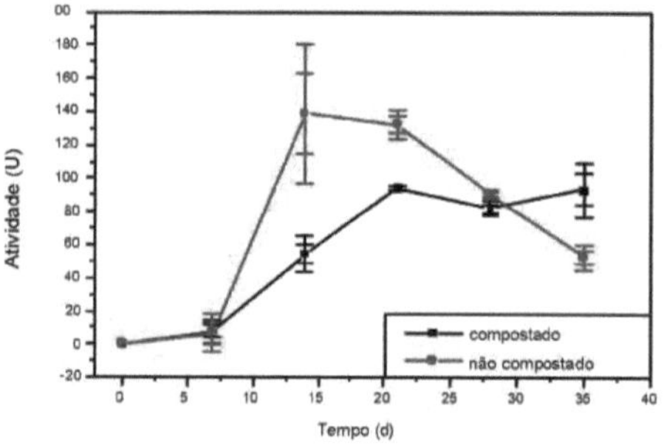


Figure 3. Laccase activity during 35 days of vegetative growth of A. brasiliensis in composted and non-composted substrate.

On the other hand, in composted substrate, maximum activity occurred later, between 21 and 35 days, in line with the profile verified by Wood & Goodenough (1977), who detected higher levels of laccase activity in A. brunnescens grown in commercial compost between 20 and 30 days of cultivation, when the fungus began to form primordia. Also in this study, the same strain under identical cultivation conditions, when the mulch layer was not added, maintained maximum levels of activity during the period studied, as was found for laccase activity in A. brasiliensis in composted substrate (WOOD & GOODENOUGH, 1977) (Figure 3).

Monitoring laccase activity and detecting high levels, especially between 20 and 30 days of growth in composted substrate, can be used by mushroom growers as an indication of the most appropriate time to lay the mulch layer, a procedure that induces fruiting and which currently has no scientifically defined schedule.

The difference in the activity of this enzyme in the two types of substrate may indicate a differentiated degradation and growth strategy. This behaviour indicates that the fungus required higher levels of laccase in the non-composted substrate, as it found the lignocellulose without any previous degradation process. This shows that the substrate preparation process, formulation and initial stage of degradation in the different substrates can be determining factors in enzyme activity levels. In non-composted substrates, there may have been limited access to lignocellulose carbon sources, hence the need to produce higher levels of laccase.

Although the individual role of enzymes in the lignocellulose degradation process is still controversial, the action of lignolytic enzymes in the biotransformation and/or biodegradation of lignin during vegetative growth is important for fungi to have access to carbon sources and promote efficient colonisation of substrates. However, excess carbon in the medium tends to suppress the induction of oxidative enzyme production for some fungi, since their production is the result of their secondary metabolism.During the period of lowest laccase activity in the non-composted substrate, after the peak in activity on the 15th day, the production of this enzyme may have been inhibited because a certain amount of carbohydrates had already been solubilised (Figure 4), i.e. after the peak in laccase activity, there was access to the carbohydrates, which were solubilised and led to an increase in their concentration in the substrate (Figures 3 and 4), and so there was a reduction in laccase activity levels until the end of the study.

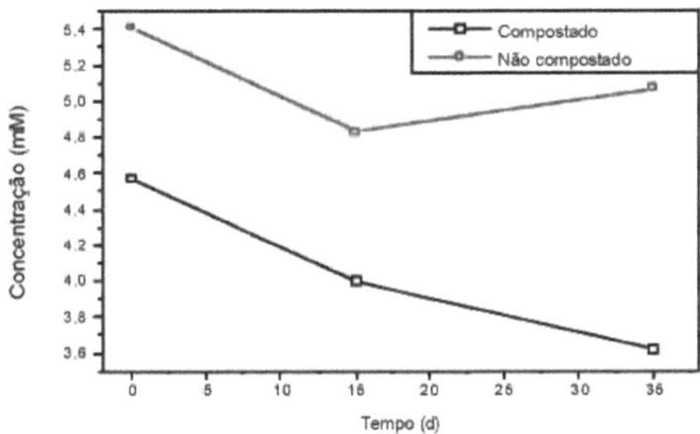


Figure 4: Consumption and initial concentration of total carbohydrates in the substrates during 35 days of vegetative growth of A. brasiliensis.

Total carbohydrates were consumed differently (Figure 4) from the 15th day onwards, and their initial concentration in the substrates was also different in the non-composted and composted substrates, 5.4 mM and

4.6 mM, respectively.

A lower initial concentration of soluble sugars was found in composted substrate, probably due to consumption during the composting process.

During the different phases of the composting process, the decomposing microorganisms consume soluble sugars and caramelise the carbohydrates, a process that eliminates water, concentrating the carbon and making the substrate suitable for consumption by the fungus. Caramelisation turns the substrate dark and selective (GERRITS, 1988). This process may be responsible for the change in the form of carbon availability in the compost.

Because it is not previously biodegraded during the preparation process, the non-composted substrate maintains the integrity of its composition and fibre structure quantitatively during sterilisation, although sterilisation can chemically alter some components.

The results obtained indicate that A. brasiliensis utilised carbon more effectively and consistently in the composted substrate, as shown in figure 5, probably due to the availability and form of the carbon in the substrate.

From the 15th day onwards, there was an increase in the concentration of sugars in the non-composted substrate. This coincides with the peak in laccase activity (Figure 3), showing that during this period, the non-composted substrate was unfavourable to the vegetative growth of A. brasiliensis, as there was some restriction in consumption. It is important to emphasise that oxidative enzymes such as laccase are part of the secondary metabolism of a group of decomposer fungi (KUES & LIU, 2000).

This dynamic of enzyme activity showed that during the first 15 days of growth in the non-composted substrate, the fungus used available carbohydrates, which did not require significant energy expenditure by the fungus to obtain. After the 15th day, it was necessary to mobilise laccase in order to access carbon to be solubilised, as shown in the soluble carbohydrate profile (Figure 4).On the other hand, in composted substrate, in order to meet the demand for carbon during vegetative growth, the fungus may have resorted to obtaining it by mobilising the enzyme xylanase, which showed high levels of activity in this substrate throughout the growth period (Figure 5).

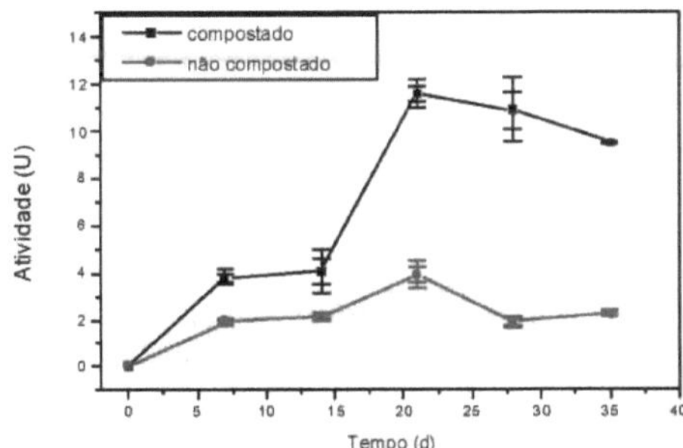


Figure 5: Xylanase activities during vegetative growth of A. brasiliensis on different substrates

The difference in xylanase activity observed between the substrates (Figure 5) may also be due to the induction of the respective gene by substances present in the composted substrate, as seen during the growth of A. brunnescens in culture medium containing compost and agar, where high levels of mRNA corresponding to the xlnA gene were detected, demonstrating the effect of induction by components of the composted substrate on the expression of this gene. These studies also showed that the fungus produced more biomass in composted substrates, which were used as a carbon source for growth.

This study showed that the gene could be repressed by the presence of monosaccharides in the growth medium (PIET et al. 1998), which are present in greater concentration in the non-composted substrate (Figure 4).

Another hypothesis that could explain the high activity of the enzyme was put forward by Iiyama et al. (1994) during an experiment with A. brunnescens on composted substrate. The author found a higher consumption of xylose compared to other sugars, probably because xylose, a derivative of xylan, was found in greater quantities in the structural formation of the cells of the component materials of the substrate used, in this case wheat straw.The activity of xylanase may have contributed to the mobilisation of carbon in the vegetative phase when the fungus grew in compost. On the other hand, in non-composted substrate, the β-glucosidase enzyme may have been responsible for carbon mobilisation, since it acts on a range of polysaccharides (CAI et al. 1998).Activity was higher in non-composted substrate throughout the study period, however, at 21 days of growth, similar levels of activity were observed, with 1297 activity units (U) in composted substrate and 1340.5 U in non-composted substrate (Figure 6).

31

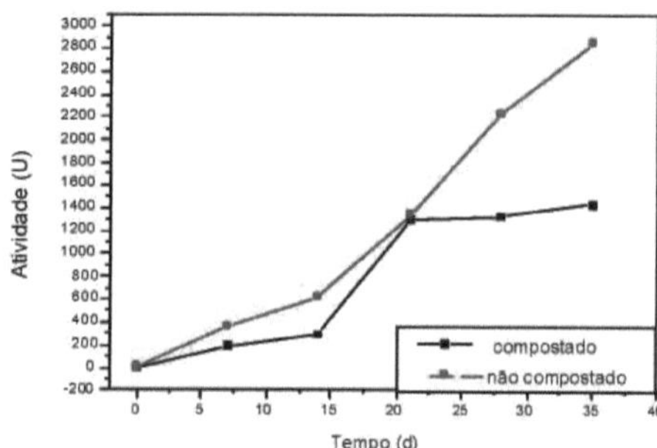

Figure 6. P-glycosidase activities during 35 days of vegetative growth of *A.brasiliensis*

Given that the concentration of soluble carbohydrates in the extract of the non-composted substrate was higher than in the composted one (Figure 4), the results obtained suggest that, quantitatively, soluble carbohydrates are not a determining factor in the production of β-glucosidase, at least in this case.

Different results were observed with Volvariella volvacea, whose β-glucosidase activity was repressed in the presence of monosaccharides, such as glucose, in the culture substrate (CAI et al. 1998). However, the enzyme's function is still unknown and studies on this genus are rare, and for some fungal species it was dispensable for mycelial growth.

The other oxidative enzyme detected during the vegetative growth of A. brasiliensis was manganese peroxidase (MnP), which showed identical activity profiles in both composted and non-composted substrates. In composted substrate, the activity of this enzyme was slightly higher from the seventh day onwards (Figure 7), remaining higher until the 35th day of vegetative growth when it reached 59.8 units of activity, 8.5% higher than the levels of activity in non-composted substrate during the same period.

MnP can act in the initial stages of high molecular weight lignin degradation (HATAKKA, 1994). However, the similarity between the profiles detected may indicate the versatility or unspecificity of this enzyme, since the reactive $Mn^+$ III that is formed favours the unspecific action of this enzyme on the substrates.The similar profile of this enzyme on both substrates did not allow us to indicate its role in each substrate, considering their differences.

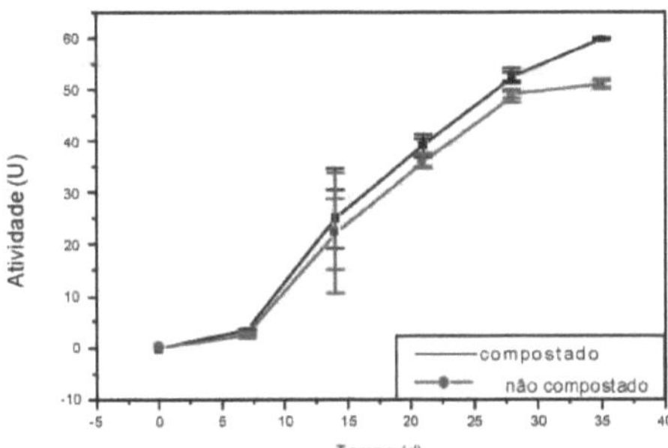


Figure 7. MnP activity during 35 days of A. brasiliensis growth in composted and non-composted substrate.

These profiles may be related to the results obtained by Bonnen et al. (1994), who detected statistically equal MnP activities in the cultivation of A. brunnescens when the species was grown in a substrate with the same formulation, subjected to different treatments, pasteurised and sterilised. However, in the present study it was not possible to make such comparisons since the substrates had different formulations.

As mentioned by various authors, the reduction in the activity of oxidative enzymes, such as laccase in particular, during vegetative growth coincides with an increase in the activity of hydrolytic enzymes required in the fruiting process. This enzyme dynamics model is the most widely accepted to explain the utilisation of lignocellulose by various fungal species (KUES & LIU, 2000; VELÁZQUEZ-CEDENO et al. 2002; WOOD & GOODENOUGH, 1977).

In order to produce mushrooms with high rates of bioconversion of lignocellulosic biomass into fungal biomass, the primary requirement is efficient colonisation of the substrate and its degradation by the mycelium, as the fungus's vegetative phase is responsible for the nutritional support of the reproductive phase (SAVOIE et al. 1996). Degradation of the substrate provides the fungus with the nutrients it needs during the production cycle, albeit in different quantitative and qualitative ways, i.e. depending on its enzyme dynamics model and the need to utilise lignocellulose, mainly as a carbon source.

Even though this study did not monitor enzymatic activity during fruiting, when high levels of hydrolytic enzyme activity occur, it is important to record the fungus's cellulolytic activity, mainly through the β-glucosidase and xylanase enzymes, since no cellulase activity was detected (FPA).

The failure to detect cellulase activity (FPA) may have been due to the stage of development that the fungus was at and/or because the substrate was not covered, one of the factors that induces the start of production of these (hydrolytic) enzymes (DURRANT et al. 1991; WOOD AND GOODENOUGH, 1977).

33

Considering the different production systems, the results obtained indicate the adaptability of A. brasiliensis to produce mycelial biomass even under unfavourable conditions, demonstrating a competitive capacity determined by the production of oxidative enzymes to degrade recalcitrant constituents and access carbon (SAVOIE et al. 1996).

### 4.3.2 Evaluation of vegetative biomass (Ergosterol and total proteins)

Differences in ergosterol content were detected between the substrates before inoculation, with values between 0.746 and 183.3 µg/g of dry substrate - non-composted and composted, respectively (Figure 8), due to the presence of microbiota characteristic of compost, which are mycelial growth promoters for A. brunnescens (STRAATSMA et al. 1993). This role in promoting mycelial growth has not been proven in the case of A. brasiliensis. However, in this study, any interaction between the promoting microorganisms and the A. brasiliensis mycelium was eliminated, as both substrates were sterilised prior to inoculation.

The amount of ergosterol produced by the mycelium is variable, depending on and/or influenced by the concentration of nutrients and the stage of development of the fungus (BARAJAS-ACEVES et al. 2002). In the present study, the concentration of ergosterol between traditional and axenic cultivation ranged from 183.3 to 501.1 and 0.746 to 217 µg/g of dry substrates, respectively, during the vegetative growth of A. brasiliensis . These results were higher than those reported by Neves (2000), who found values varying between 163.62 and 187.80 µg/g of dry compost. However, the author reports ergosterol content of around 618 µg/g of composted substrate before inoculation, unlike the values found in the present study.

The total carbohydrate content of each substrate was different at the start of growth, with higher values detected in the non-composted substrate. However, this apparent availability of carbon was not enough to promote constant vegetative growth in these conditions during the monitored period. It is important to emphasise that during the initial 15 days of vegetative growth of A. brasiliensis there was a similar rate of biomass production between the substrates (Figure 8). On the other hand, the presence of higher concentrations of soluble carbohydrates in the non-composted substrate may have suppressed lignocellulosic degradation, as suggested by Hatakka (1994).

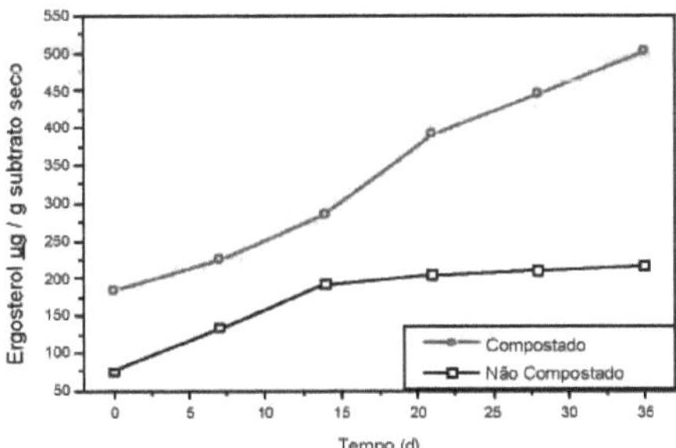


Figure 8. Ergosterol content in the mycelium of A. brasiliensis, assessed during 35 days of vegetative growth in composted and non-composted substrate.

The available soluble carbohydrates were consumed and promoted a similar increase in growth up to the 15th day in both substrates, however, in the non-composted substrate, from the 15th day onwards, the fungus may have resorted to degrading lignocellulose as a carbon source, since the soluble carbon had been consumed. The activation of secondary metabolism by the production of oxidative enzymes responsible for the degradation of lignin showed that the carbon sources were not adequately available in this substrate, since secondary decomposer fungi have a preference for carbon sources that have undergone "fermentation" as is the case in composted substrates (PIET et al. 1998). In these substrates, carbohydrate consumption and biomass production were therefore more efficient.

The need for energy consumption through the synthesis of extracellular proteins, mainly laccase, to mobilise carbon in the non-composted substrate led to a reduction in vegetative growth, accompanied by the ergosterol content in relation to the composted substrate (Figure 8), although the protein synthesis profiles were similar during the study period (Figure 9).

The similarity between the protein biosynthesis profiles observed between the substrates, albeit at different levels, indicates that even in conditions less favourable to growth, as seen through the laccase activity profile in non-composted substrate, A. brasiliensis has not altered its protein synthesis metabolism, a fact that suggests its versatility for cultivation in conditions different from those used in traditional cultivation, based on composted and pasteurised substrates. Some indirect methods of evaluating fungal mycelial biomass, such as the dosage of soluble proteins or the evaluation of extracellular laccase activity, have been correlated with vegetative biomass, evaluated by determining dry weight and ergosterol content. In the case of the edible fungus Lentinula edodes, or shiitake, grown on solid media, a correlation was found between biomass

35

production, measured by ergosterol concentration, and the concentration of total soluble proteins (OKEKE et al. 1994).

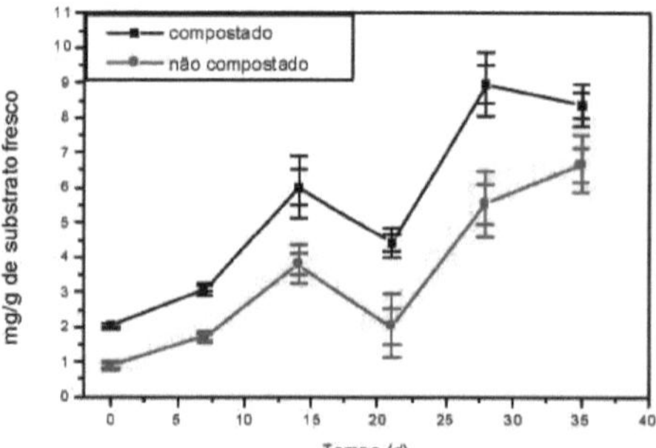


Figure 9. Total protein content of A. brasiliensis during vegetative growth in traditional (composted) and axenic (non-composted) cultivation.

During the vegetative growth of A. brasiliensis, the laccase activity in composted substrate and the total protein content showed a significant correlation in relation to the vegetative biomass evaluated by determining the ergosterol concentration, with r =0.93 (<0.01) and r = 0.87 (<0.05), respectively. These results indicate that the laccase enzyme in particular could be used to assess the production of mycelial biomass of A. brasiliensis in composted substrate. This conclusion was also reached for A. brunnescens when grown in commercial compost (MATCHAM et al. 1985).

In the non-composted substrate, the fungus showed differences in metabolism, probably due to the different availability of nutrients. These changes influenced both parameters, so that in the non-composted substrate there were no significant correlations between laccase and total protein synthesis and laccase and ergosterol, with r = 0.74 (< 0.10) and r = 0.77 (< 0.10), respectively.

Studies carried out on A. brunnescens in solid media showed a significant correlation between phenol oxidase laccase activity and biomass, assessed both by determining dry weight and ergosterol concentration. However, the correlation was limited to a period of 30 days. A significant correlation between these parameters and the concentration of total proteins was also observed (MATCHAM et al. 1985).

The results obtained in the non-composted substrate showed a response to the unfavourable condition for A. brasiliensis, which had an impact on lower mycelial growth, although the substrate in the growing bags was completely colonised by denser mycelium on the outside. Despite the different metabolism strategies depending on the cultivation substrate, strain UFSC-51 has the potential to be used in axenic cultivation,

36

depending on future experiments in the fruiting phase.

As for protein concentration, the composted substrate initially showed a higher level than the non-composted substrate. Protein biosynthesis in both conditions (Figure 9) showed identical profiles over time, suggesting that under different growth conditions, A. brasiliensis maintained a similar profile of biosynthetic activity. Although the results indicate that the non-composted substrate is an unfavourable environment for the growth of the fungus, it can be seen that A. brasiliensis (UFSC-51) was versatile enough to grow in different types of cultivation processes, composted and non-composted.

## 4.4  CONCLUSIONS

- Agaricus brasiliensis produced the same enzymes (lacase and Mn peroxidase - oxidative and xylanase and â-glucosidase - hydrolytic) in the different substrates, without producing LiP (lignin peroxidase) and cellulase (Cel - FPA) in the substrates;

- The substrate preparation process and its stage of degradation changed the availability of nutrients and altered the enzyme dynamics;

- Supplemented substrates based on royal palm leaves showed viability for use in the axenic cultivation of A. brasiliensis ;

- A. brasiliensis showed greater vegetative growth, as assessed by ergosterol, in composted substrate;

- The composted substrate was more suitable for vegetative growth;

- The axenic substrate led to lower mycelial growth under the formulated conditions;

- The fungus maintained an identical protein biosynthesis profile even when cultivated under different conditions;

- Indirect methods traditionally used to monitor vegetative growth (Lacase and total proteins) can be used for A. brasiliensis when grown in composted substrate;

# CHAPTER 3

**5.** Changes in lignocellulosic components and acidification during vegetative growth of Agaricus brasiliensis on different substrates


### Summary
Several commercially cultivated fungi are produced on previously processed agro-industrial waste. Information on the utilisation of this waste for conversion into vegetative biomass provides subsidies for improving conversion efficiency in different substrates. In this chapter, the degradative capacity of Agaricus brasiliensis was evaluated through the changes that occurred in sterilised substrates for axenic and traditional cultivation (compost). Lignocellulose degradation was monitored during 35 days of vegetative growth by analysing Klason lignin, infrared spectroscopy (FTIR), pH and C/N ratio. The relative concentration of lignin increased in composted substrate and decreased in non-composted substrate. The infrared spectra showed that the non-composted substrate had high initial concentrations of the groups responsible for the band at 1650 cm$^{-1}$. Polysaccharides (1110cm$^{-1}$) were consumed more efficiently in compost. In the non-composted substrate, a band appeared at 1325 cm$^{-1}$, related to laccase activity. The pH variations showed a tendency towards acidification in both substrates. The C/N ratio decreased in the compost and increased in the axenic substrate. The results obtained showed a more pronounced consumption of polysaccharides in compost; however, the main structural changes occurred in sterilised substrates as a result of laccase activity.

Keywords: FTIR, lignin, compost, vegetative growth, Agaricus brasiliensis , biodegradation


## 5.  1INTRODUCTION

In order to utilise carbon from lignocellulosic substrates for growth, both in the vegetative and reproductive phases, basidiomycetes belonging to the white rot group, known as "White-rot", degrade the basic components of the plant cell wall, which include cellulose, hemicellulose and lignin, unlike the brown rot or "Brawn-rot" fungi, which selectively degrade polysaccharides (PANDEY & PITMAN, 2003). Carbon utilisation by these fungi, such as Agaricus, Pleurotus and others, varies according to the stage of development they are in, i.e. during vegetative growth, consumption is aimed at forming new hyphae and during reproductive growth at forming fruiting, showing a different nutritional demand for each stage (DURRANT et al. 1991; MOL, 1989).

Durrant et al. (1991) showed that during the life cycle, under cultivation conditions, Agaricus brunnescens (champignon) degraded some lignin fractions, preferentially during vegetative growth, and the polysaccharides were utilised during fruit formation.

For there to be efficient use of the substrate's carbon for growth in the vegetative and reproductive phases, especially when the fungus is produced at commercial level, it is essential that the substrate has an adequate chemical and physical balance and, in the case of pasteurised composted substrates, also a microbiological balance.

Certain parameters are often used to monitor the degradation process, allowing changes and/or modifications to be detected in certain substrates, mainly as a result of the action of extracellular enzymes produced by the fungi. Among the most important parameters that allow the nutritional balance of the substrate to be assessed is the ratio between the carbon and nitrogen content present in the growing substrate. The optimum C/N ratio is selected during the initial formulation of the substrate and determines the quality of the final product, both for composted and axenic substrates (GERRITS, 1988). In these two production processes, carbon is supplied to the fungus by the bulky component of the substrate, i.e. the straw, and nitrogen by supplementary elements such as various types of manure and bran, as well as that present in the biomass formed during the composting process (EIRA, 2000; GERRITS, 1988).

However, in non-composted substrates there are less drastic physical-chemical, microbiological and particularly structural changes during preparation, unlike what happens in the composting process. In these substrates, the basic components are not modified by microbiological activity, while in composted substrates this change occurs, thus leading to greater structural and chemical complexity, as can be seen, for example, in the complexing of proteins with the lignin polymer (GERRITS, 1988).

If there is a quantitative nutritional imbalance of the components in the substrates, metabolic changes can be observed in the fungus and its degradative capacity, as well as promoting the development of competing organisms in both production systems (KEREM et al. 1992; GERRITS, 1988).

In the case of composted substrates, nutritional balance is fundamental for selectivity, so that microbiological succession during the process between fungi and bacteria, including actinomycetes, can be established by chemically altering the constituents of the compost, resulting in a series of modifications, including an increase in microbial biomass, which serves as a source of nutrients for the fungus of interest. In some situations, this biomass is responsible for promoting the growth rate of the fungal mycelium, conferring selectivity to the compost and inhibiting the development of competitors that reduce productivity (FERMOR et al. 1985 apud IIYAMA et al. 1994).

Part of the polysaccharides present in the initial compost formulation is degraded by microorganisms during composting, and only 17% is effectively utilised by A. brunnescens during the production cycle (IIYAMA et al. 1994). Nitrogen, which is initially present in the form of proteins, is metabolised during composting, generating ammonia ($NH_3$) in a process known as ammonification. The ammonia is fixed by microorganisms together with the lignin, forming the lignin-humus complex, which is then used as a source of nitrogen by the fungus. The transformation of nitrogen, the consumption of polysaccharides and microbial respiration lead to a reduction in the final C/N ratio, both during composting and during the cultivation of Agaricus (GERRITS, 1988).

In Brazil, the majority of compost produced has a C/N ratio of around 17/1 at the end of composting, most of which is used to grow A. brunnescens (BRAGA & EIRA, 1998). On the other hand, there are no known producers who use axenic substrates as an alternative for growing this species, which could be an important strategy for optimising productivity.

Recently, it was found that A. brasiliensis required substrates with a C/N ratio of around 37/1 to obtain higher yields,
different from those traditionally used in the country (KOPYTOWSKI FILHO, 2002).

The chemical, physical, microbiological and structural changes that occur during compost production are similar to those that occur during the growth phase of the fungus of interest. The reduction in the C/N ratio and polysaccharide content during composting also occurs during the growth of the fungi in the composted substrates. In the hydrolysis of polysaccharides, organic acids are generated that promote a reduction in the hydrogenionic potential (pH) of these substrates (SCHISLER, 1982). During the degradation of lignin, acid derivatives are generated which can also contribute to acidification. Bonnen et al. (1994), during the production of A. brunnescens in compost, observed a slight reduction in pH, which ultimately remained at around 6.0 (5.8-6.2).

Monitoring the degradation of lignin and/or its modification using enzymatic, chemical and spectroscopic methodologies is well represented in the literature (IIYAMA et al. 1994; SOARES & DÚRAN, 2001; TOUMELA, 2000). This monitoring can be carried out using different methodologies, and traditional chemical methods such as the determination of Klason lignin can be used to quantify it. Alterations in the structure of the lignin molecule, determined either chemically or enzymatically, can have implications for the interpretation of evaluations (IIYAMA et al. 1994; TUOMELA, 2000).

Some authors therefore consider that there may be an increase in the proportion of lignin during the growth of wood-degrading fungi, due to the preferential consumption of other elements, while the lignin content is maintained, even if modified by oxidation, thus resulting in an increase in the relative quantity (IIYAMA et al. 1994). On the other hand, if this polymer is degraded or transformed, classical analytical methodologies such as klason would indicate a reduction in its relative concentration and complementary methodologies would show these changes.

Classical methodologies are therefore being replaced or complemented by less costly procedures that use smaller quantities of substrate and produce results more quickly (CHEN et al. 2000; PANDEY & PITMAN, 2003; TUOMELA et al. 2000).

Infrared spectroscopy (FTIR) is one of the techniques used to study the chemical degradation of wood, which requires minimal sample preparation and small quantities, unlike traditional gravimetric methods, where larger samples are required (PANDEY & PITMAN, 2003), serving as a qualitative tool to characterise chemical groups present in organic matter without destruction (INBAR et al. 1989). The FTIR technique refers to the study of the interactions of infrared light with matter, where the frequency of the waves is inversely proportional to the energy. This technique provides information on functional groups, molecular geometry and intra- and inter-molecular interactions, where the frequency at which absorption occurs can indicate the type of functional group present in the substance (www.nelsonlabs.com).

Analyses of FTIR spectra using not only peak frequency but also relative intensities provide information on qualitative changes, even though polymers can undergo chemical alterations without destroying cell morphology (GIVEN et al. 1984).

The FTIR spectra generated represent a complex set with cellulose C-O-C bonds vibrating centred at $1060\text{-}1100$ cm$^{-1}$ . The spectra of hemicellulose are very similar to those of cellulose. In general, the region around $1110$ cm$^{-1}$ can be used to evaluate polysaccharides (GIVEN et al. 1984).

Lignin in turn shows bands at $1460$ cm$^{-1}$ represented by the vibration of aliphatic groups, but its oxidation is accompanied by bands represented by aromatic groups at $1650$ cm$^{-1}$ and $1325$ cm$^{-1}$ , as a result of the action of phenoloxidase enzymes such as lacase (GIVEN et al. 1984; SOARES & DURAN, 2001).

Chen et al. (2000), in a study aimed at mapping changes in compound composition during the growth of A. brunnescens , showed that there was an increase in the region of aromatic structures represented at $1650$ cm$^{-1}$ and in the band in the region of $1325$ cm$^{-1}$ , representing groups that appear due to changes in lignin. The authors also found a notable reduction in polysaccharides (region around $1110$ cm )$^{-1}$

The objectives of this work were to monitor biodegradation-related parameters for composted and non-composted substrates in the cultivation of A. brasiliensis during the vegetative growth phase, using traditional chemical techniques complemented by spectroscopy techniques. These studies were subsequently integrated with enzymatic studies (Chap.II) in order to establish a degradation model for A. brasiliensis during vegetative growth by comparing its growth on two types of substrate (composted and axenic).

## 5.2  MATERIALS AND METHODS

The stages of substrate preparation, inoculation and sampling, as well as the fungal strain used to obtain the results presented in this chapter, follow the previous description of the materials and methods in Chapter II, plus additional analysis methodologies.

### 5.2.1  Evaluation of Klason lignin

500 mg of each previously dried substrate was weighed out and then placed in 250 mL screw-top glasses. 5mL of 72% $H_2SO_4$ was added, i.e. 1mL for every 100 mg of sample. The mixture was placed in a water bath at $30\pm0.5°C$ and stirred frequently. After 1 hour, the sample was diluted using 28 mL of distilled water for every 1 mL of sulphuric acid and transferred to 125 mL beakers. Secondary hydrolysis was carried out in an autoclave at 121°C for 1 hour. The still hot solution was filtered through pre-weighed filter paper under vacuum pump suction and the residue washed with a known quantity of hot $H_2O$ to remove the acid. The filter paper discs containing the residue from the washed samples were dried at 65°C for 72 hours. They were then weighed and the mass obtained was calculated according to the difference between the initial and final weight of the discs.

### 5.2.2  Infrared spectroscopy analyses - FTIR

The analyses were carried out using infrared spectroscopy on an FTLA 2000 apparatus, using 400 scans/readings per minute between 400 and 4000 **cm-1. The spectrometer was equipped with software to perform the "Fourier** Transform" carried out at the Analysis Centre of the Chemistry Department, Federal University of Santa Catarina.

To do this, tablets were made containing 99.9 mg of KBr and 0.1 mg of previously dried, ground and sieved (1 mm) substrate sample. The tablet was made after macerating these two components with a pestle and pestle. The resulting powder was placed in a suitable holder and pressed under a pressure of 10 tonnes for 1 minute. The tablet was removed from the support and placed directly into the apparatus for reading.

Due to the heterogeneity of the different substrates, the reproducibility of the analysis method and the sampling method (Chapter II) were evaluated to validate the methodology by taking four sub-samples from the growing bags (sampling units) as described in Chapter 2 - Sampling, and submitted to FTIR spectroscopy individually.

### 5.2.3   pH analysis

The pH was determined in triplicate by mixing the fresh substrate (10g) with water (50 mL) and after 30 minutes the supernatant was analysed (TEDESCO et al. 1985).

### 5.2.4   Elemental analyses

The elemental analyses of each substrate with regard to the parameters carbon, nitrogen and organic matter over time were carried out at the EPAGRI laboratory in Caçador, SC, numbered 4140 to 4146.

## 5.3 RESULTS      AND DISCUSSION
5.3.1   Determination of lignin concentration (Klason method)

The initial lignin concentration (day zero), assessed in both substrates using the Klason methodology, was higher in the non-composted substrate compared to the composted one, 27.0 and 24.4 mg/g of dry substrate, respectively (Figure 10). This can be explained by the preparation process of each substrate and the different formulation, taking into account the differences in the lignin content of the ingredients in each formulation. One of the substrates included maize straw and corncobs, while the other included royal palm straw and sawdust. In the first case, composted substrate, around 60 per cent of lignocellulosic ingredients were included in the fungus inoculation phase, while in the second case, these ingredients accounted for around 80 per cent (Substrate preparation - chapter II).

The results show a 9.9 per cent increase in the relative lignin content at the end of 35 days of growth of A. brasiliensis in composted substrate (24.4 to 26.8 mg/g dry substrate). However, with three sampling points, an increase in relative lignin content was identified on the 15th day of growth (31.7 mg/g dry substrate) (Figure 10). The increase in lignin content was also observed by Iiyama et al. (1994) during the production of A. brunnescens . Inbar et al. (1989) also reported such an increase during the composting process. It is important to emphasise that the greatest increase in lignin concentration was detected during composting (6.1%), which is higher than that observed during the species' cultivation cycle (0.6%).

One of the reasons pointed out by various authors for this increase in certain substrates undergoing degradation may be the recalcitrant nature of the polymer, resulting from its structural and chemical complexity and consequent low biodegradability. In other words, the increase in lignin content is not due to a process of biosynthesis, but rather to the preferential reduction of other components, thus increasing its relative quantity (IIYAMA et al. 1994).

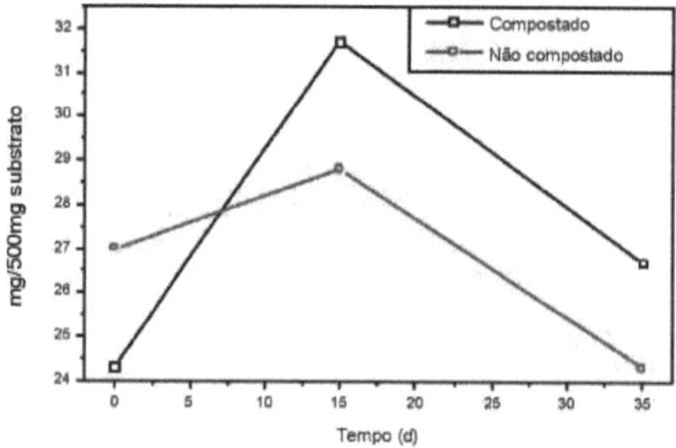


Figure 10. Variations in relative lignin content during 35 days of vegetative growth of A. brasiliensis in composted and non-composted substrate

Reinforcing this hypothesis, the consumption of soluble carbohydrates by the fungus occurred more efficiently in composted substrate, as shown in Figure 4 (Chapter II), which may have led to an increase in the relative lignin content in this substrate. In addition, a lower level of lacase activity was observed in composted substrate during 35 days of vegetative growth of A. brasiliensis , an enzyme involved in lignin degradation.

Lacase, as already mentioned, can provide access to the carbon source or take part in the process of removing potentially toxic intermediate metabolites present in the medium, i.e. it participates in both the degradation and/or biotransformation of lignin and the detoxification of the growth medium from the by-products of such degradation (SOARES & DÚRAN, 2001).

Lower levels of laccase activity in composted substrate suggest greater carbon availability during the 35 days of vegetative growth, with a slight increase in activity from 21 days onwards, but this did not mean changes in soluble carbohydrate content. Peaks in laccase activity between 20 and 30 days have also been reported during the vegetative growth of A. brunnescens (WOOD & GOODENOUGH, 1977). The presence of carbon at levels suitable for the species or in excess in the growth medium may be responsible for the repression in lignolytic activity (KEREM et al. 1992).

In non-composted substrate, there was a reduction of approximately 10 per cent in relation to the initial lignin content at the end of 35 days, from 27.0 to 24.4 mg/g of dry substrate. The reduction in lignin content may have been due to the carbon limitation of this substrate, since laccase, which has been identified as a player in the degradation of lignin, participates in the secondary metabolism of fungi under conditions such as carbon limitation, showing high levels of enzyme activity in this substrate. This can be seen in figure 3 in chapter II.

The initial consumption of soluble carbohydrates in the non-composted substrate was similar to the composted substrate and may have promoted a slight increase in the content of vegetative biomass in this substrate until the 15th day, when the fungus is thought to have already consumed the readily assimilated carbohydrates available. In order to continue its vegetative growth, it mobilised oxidative lignolytic enzymes, mainly lacase, to access the carbon sources in the lignocellulose. For this reason, vegetative growth, as assessed by the ergosterol content, was compromised from this date onwards (Figure 8 - Chapter II), showing the need for energy expenditure for this metabolic process.At the end of 35 days of vegetative growth, the lignin concentration in the non-composted substrate was close to that seen in the composted substrate before the fungus grew (day zero).

### 5.3.2  Infrared spectroscopy

One of the problems that could be encountered in this type of analysis is the possible low reproducibility of the sampling method (Chapter II), due to the heterogeneity of the composition of the substrates, especially non-composted substrates. To check reproducibility, four sub-samples were taken from the same sampling unit and subjected to spectroscopy (FTIR). The spectra obtained were highly similar, thus indicating reproducibility.

The mycelium, when mixed in different concentrations with the non-colonised substrates (Data not shown), contributed proportionally to the absorbances at 1650 cm$^{-1}$ , 1460 cm$^{-1}$ and 1100 cm$^{-1}$ . Therefore, as discussed in the following sections, the ratio between the absorbances of the bands will be used to make it easier to visualise and understand these changes. The changes between the ratios of the relative absorbance intensities observed during the fungus's vegetative growth are the result of effective chemical changes in the various functional groups of the substrates, caused by the fungus's enzymatic action. During the period studied, 35 days, three spectra were generated per treatment, however, only two are presented for the purpose of comparison between the initial point, the substrate not colonised by the fungus, and the substrate after 35 days of vegetative growth.

5.3.2.1  Spectral differences   between the substrates prior to fungal growth.

According to Figure 11, spectral differences were observed between the intensities in the region of 1460cm$^{-1}$ and 1650cm$^{-1}$ , evidenced by the ratio between the absorbances 1650/1460cm$^{-1}$ of 0.68 and 1.26, respectively, for the spectra of the samples of non-composted and composted substrates (Table 2).

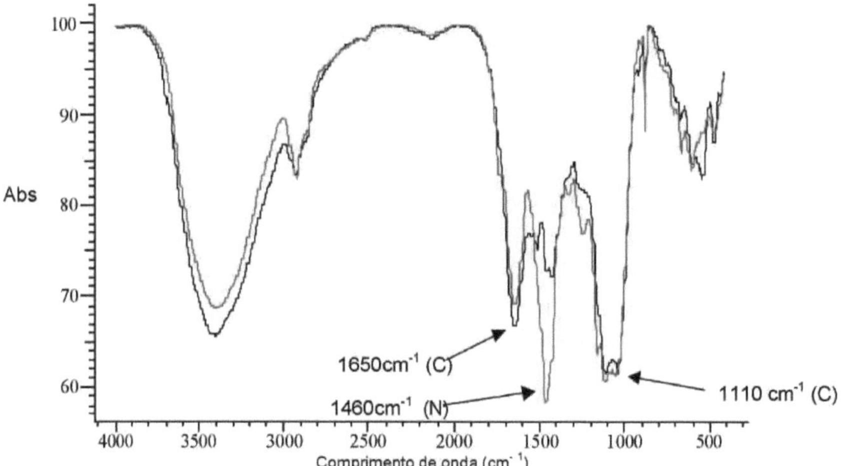


Figure 11. FTIR spectra showing initial differences between composted (N - black) and non-composted (C - red) sterilised substrates prior to colonisation.

The composted substrate had a significantly higher content of aliphatic groups responsible for absorption at 1460cm"₁. The absorption bands in this region may be produced by the action of microbial enzymes (SHARMA & KILPATRICK, 2000). It is important to point out that the composted substrate requires intense microbiological activity during its preparation.

Table 2. Quotients between the abs orbances of the main bands observed in the FTIR spectra of the samples obtained before (initial) and after (35 days) the vegetative growth of A. brasiliensis (final) in the respective composted and non-composted substrates.

| Ratio between abs. | Composted | | Not composted | |
|---|---|---|---|---|
| | Home | Final | Home | Final |
| 1650/1460 | 1,26 | 2,30    (2x) | 0,68 | 2,83 (~4x) |
| 1650/1100 | 0,82 | 1,23 (1,5x) | 0,73 | 1,62 (~2x) |
| 1460/1100 | 0,53 | 0,64 (1,15x) | 1,07 | 0,57 (~-2x) |
| 1100/610 | 2,85 | 2,20 | - | |
| 1100/2925 | - | - | 2,69 | 3,05 |

5.3.2.1. Structural changes in composted substrate after 35 days of degradation by A. brasiliensis .

The main spectral changes in composted and pasteurised substrate are shown in Figure 12. The

46

values of the ratios between the absorbance intensities of the various bands are shown in Table 2. The increase in the respective ratio for the bands in the 1650/1460 cm regions[-1] is clearly shown by the values for the initial and final points, 1.26 and 2.30.

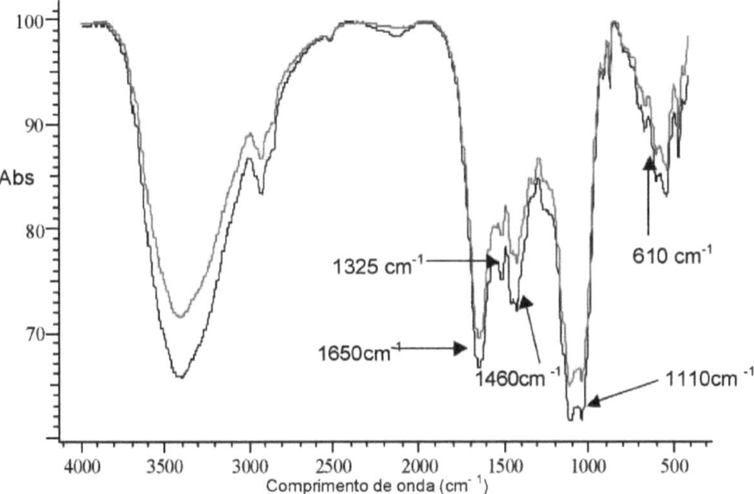


Figure 12. FTIR spectra of composted and pasteurised substrate during vegetative growth of Agaricus brasiliensis Initial (Black) and Final (Red).

At 1650 cm[-1] , a band that may refer to carbonyl groups, there was an increase of around 50% (Table 2) during the vegetative growth period of A. brasiliensis (35 days), a result also verified by Chen et al. (2000) during the growth of A. brunnescens. In addition, other studies with basidiomycete fungi relating to the degradation of lignin and its derivatives, including those present in industrial effluents, have also demonstrated the same type of chemical modification. In these cases, this type of alteration has been related to the action of lignolytic enzymes, such as laccase (SOARES & DÚRAN, 1995).

On the other hand, the appearance of the band in the 1325 cm region[-1] , in the spectrum of the final sample (after fungal growth), may be related to the presence of
syringyl groups, typical of the lignin structure. This evidence, if confirmed, may indicate an increase in the relative content of these chemical groups, i.e. it represents a preferential degradation of other component groups of the lignin structure such as guaiacyl groups, possibly through the action of laccase (FAIX et al. 1985).

At this point, it is interesting to emphasise the need to carry out other analyses using different

47

methodologies, such as pyrolysis followed by gas chromatography/mass spectroscopy, in order to complement the information needed to unequivocally identify the presence of these groups, as done by Chen et al. (2000).

Otherwise, this behaviour could be explained by the preferential degradation of carbohydrates over lignin, as also observed by Chen et al. (2000). These results are in agreement with those obtained from the chemical dosage of lignin in the composted substrate, as described above (section 5.3.1).

The reduction in the intensity of the bands in the 1100 cm region$^{-1}$ was due to the consumption of carbohydrates (Figure 12). This is shown by the reduction in the ratio between the 1100 cm$^{-1}$ and 610 cm$^{-1}$ bands from 2.85 to 2.20, respectively. It should also be noted that the band at 610 cm$^{-1}$ was used to calculate this ratio because this band did not vary in both spectra (initial and final).

5.3.2.3 Structural changes in the non-composted substrate subjected to 35 days of degradation by A. brasiliensis

In non-composted and sterilised substrate, comparisons were made taking into account the initial and final spectra (Figure 13). A comparison of the ratios between the absorbance intensities of the respective bands (Table 2) showed an increase in intensity in the 1650 cm region$^{-1}$ relating to aromatic groups. For this substrate, the considerations made earlier for bands at this intensity for composted substrate (previous section) are also valid, as well as reinforcing the arguments used earlier about changes due to high laccase activity in non-composted substrate, since in this case, the levels of laccase activity observed are even higher. Consistent with this statement is the fact that there was a sharp increase in the ratio between the absorbance values for the 1650/1460 cm bands$^{-1}$ , between the initial and final samples of the fungus' vegetative growth, of 0.68 and 2.83 respectively.

There was also a reduction in the region of 1460 cm$^{-1}$ , due to a probable decrease in the relative content of aliphatic groups, as well as a significant increase in the relative intensity of the band at 1325 cm$^{-1}$ , which, as already mentioned, may be related to the syringyl groups of the lignin side chains. However, there is a need to use more efficient analytical methods to prove the preferential degradation of certain lignin polymer structures. On the other hand, it can be confirmed that the changes that contributed to the appearance of the band at 1325 cm$^{-1}$ were due to laccase activity (SOARES & DÚRAN, 1995).

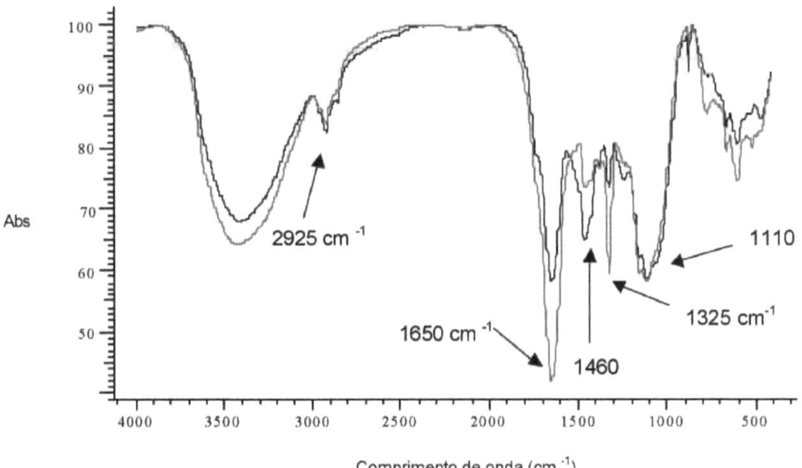


Figure 13. FTIR spectra of non-composted and sterilised substrate during vegetative growth of Agaricus brasiliensis Initial (black) and Final (Red).

With regard to carbohydrates, represented by the 1100 cm region[-1] , it was noted that the bands did not show any significant changes (Figure 13). This is shown in Table 2, where the ratios between the 1100/2925 cm[-1] bands are presented, which show a slight increase (from 2.69 to 3.05 at the beginning and end, respectively), showing that there was no actual consumption of carbohydrates, but rather a slight increase in their relative quantities, probably due to the shape of these polymers, a fact that accentuated energy consumption by the fungus. This confirmed the fungus' "preference" for composted substrates. It should be emphasised that the band in the region of 2925 cm[-1] was used to calculate the ratio between 1100/2925 cm[-1] because it was constant in both spectra (initial and final).

### 5.3.3 Acidification of the substrates (Hydrogen Potential - pH)

The pH variations in the composted and non-composted substrates showed similar profiles, with decreasing values over the course of vegetative growth. The changes were in values from 6.9 to 5.4 in the composted substrate and 7.35 to 6.1 in the non-composted substrate (Figure 14), i.e. there was significant acidification of the growing substrates.

This behaviour was different to that observed by Bonnen et al. (1994) during the growth of A. brunnescens in composted substrate, where pH values remained around 6.0 during the cultivation cycle, varying from 5.8 to 6.2, i.e. the pH variation was discrete. More effective variations in pH levels in relation to A. brunnescens

49

may indicate a difference in degradative capacity between species and/or strains, since such profiles can be attributed to the generation of organic acids, resulting from the degradation of sugars consumed as a carbon source for vegetative growth, although the oxidation of lignin which leads to the formation of acid derivatives cannot be ignored (SCHISLER, 1982).

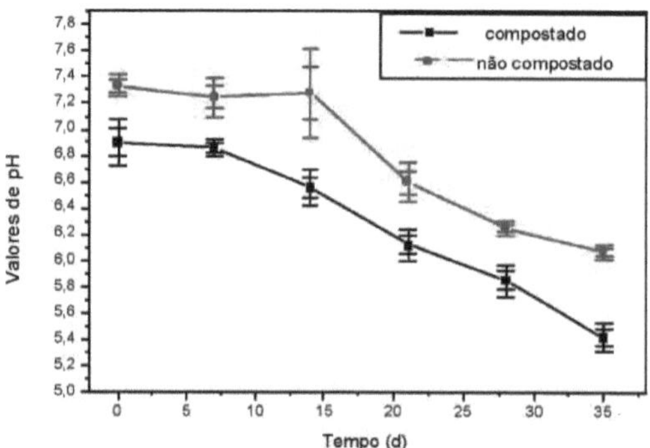


Figure 14. pH values during vegetative growth of A. brasiliensis in composted and non-composted substrate over 35 days.

In this context, the addition of correctives (gypsum and $CaCO_3$) during the preparation processes of both substrates provides a better texture, preventing the formation of free water, as well as mitigating the acidification of the medium, through the buffering role played by $CaCO_3$.

### 5.3.4 Elemental analyses

Analyses of the carbon and nitrogen content in both substrates during the vegetative growth of A. brasiliensis showed that the C/N ratio after 35 days in the composted substrate decreased from 20/1 to 17/1 and in the non-composted substrate it increased from 25/1 to 27/1 (Table 3). Carbon consumption shows that obtaining nutrients in the composted substrate was more efficient in this condition (Table 3) and was responsible for the reduction in the C/N ratio. This statement can also be supported by the formation of greater biomass in this substrate.

Table 3. Changes in the composition of carbon, nitrogen and organic matter in composted and non-composted substrate during 35 days of Agaricus brasiliensis growth.

| Elements / time | | Home | 15 days | 35 days |
|---|---|---|---|---|
| **Carbon** | Composted | 37.2 | 35.9 | 35 |
| % | Not composted | 44.7 | 44.2 | 44.3 |
| **Nitrogen** | Composted | 1.81 | 1.7 | 1.97 |
| % | Not composted | 1.76 | 1.48 | 1.63 |
| **Organic Matter** | Composted | 66,96 | 64,62 | 63 |
| % | Not composted | 80,46 | 79,56 | 79,74 |
| **C/N ratio** | Composted | 20/1 | - | 17/1 |

|  | Not composted | 25/1 | - | 27/1 |

With regard to nitrogen content, the amount consumed may have been effectively used to form extracellular protein, unlike the composted substrate, where there was an increase in biomass and a reduction in the C/N ratio, showing more effective carbon consumption (Table 3) throughout the study period. These results shed light on the reduction in the relative concentration of lignin (Figure 10) and the decrease in the C/N ratio (Table 3), i.e. when there is more effective consumption of nitrogen. Notably, the composted substrate consumed higher proportions of organic matter, a fact that supports the increase in relative lignin content and greater biomass formation. However, other analyses, such as the ash content, would need to be carried out in order to show the constitution of these materials more clearly.

## 5.   4CONCLUSIONS

-   The cultivation of A. brasiliensis during vegetative growth increased the relative lignin content in the composted substrate as a result of the preferential consumption of other more easily degradable constituents. In non-composted substrate, the lignin content decreased, possibly due to greater accessibility to easily degradable polysaccharides.

-   The C/N ratio of the composted substrate decreased, unlike the non-composted substrate, where the ratio increased.

-   Polysaccharides are most consumed by A. brasiliensis during cultivation in the composted substrate.

-   The non-composted substrate initially had a higher concentration of aromatic substances than the composted substrate.

-   In both substrates there was an increase in the content of aromatic substances ($1650$ cm$^{-1}$ ) and a reduction in the content of aliphatic substances ($1460$ cm )$^{-1}$

-   In composted and non-composted substrates, the pH tended to acidify.

-   Organic matter consumption was higher in the composted substrate than in the non-composted substrate, implying greater biomass formation.

### 6. FINAL CONSIDERATIONS

Agaricus brasiliensis has the characteristic enzymatic profile of a white rot fungus, organisms capable of degrading lignin. However, this ability proved to be optional depending on the composition of each substrate. When the fungus grew on a substrate that made carbon available in the most accessible form, composted substrates, there was selective degradation of these, to the detriment of lignin degradation. The selective enzymatic action on sugars and organic matter, in relation to other polymers, led to an increase in the relative content of lignin in relation to the total material. This contributed to a reduction in the ratio between carbon and nitrogen.

On the other hand, when the carbon was not in the most accessible form, in the case of the non-composted substrate, the fungus mobilised lignolytic enzymes (mainly laccase, especially from day 15 onwards) which required greater energy expenditure for their biosynthesis. These enzymes, as well as providing access to more suitable carbon sources, led to a reduction in the relative amount of lignin in the substrate. As a result, there was a more effective consumption of nitrogen, an increase in the C/N ratio and a reduction in mycelial growth. For these reasons, biomass production in this substrate was lower than in the composted substrate.

The composted substrate therefore provided more suitable conditions for the vegetative growth of A. brasiliensis. The non-composted substrate had the lowest vegetative biomass, a fact that does not rule out the potential of this strategy as an alternative for growing A brasiliensis. However, the formulation used in this study, particularly the C/N ratio and the chemical composition of the carbon components, were not suitable for the vegetative growth of A. brasiliensis.

However, several strategies could be tried to make the use of formulations based on royal palm in axenic cultivation viable. The first consists of adjusting the formulation according to the C/N ratio, with more suitable nitrogen supplements, since various studies have shown the potential of highly supplemented axenic substrates to obtain higher yields.

The second possibility would involve treating the bulky material, especially that with a high lignin content, a factor that restricts carbon sources, with prior chemical or biological treatments in order to cause partial de-lignification. In the case of considering the biological route, it would be feasible to grow A. brasiliensis on the residual substrate of Pleurotus , a fungus with a high capacity for degrading lignin, as well as being an edible and medicinal fungus of commercial interest. Finally, and as a third alternative, it would be advisable to pre-ferment the bulky material prior to the sterilisation process. In this way, the partial degradation of the lignin by the native microbiota could be facilitated, thus subjecting this material to a traditional, but short-lived, composting process.

However, before using any type of substrate at a commercial level, we must emphasise the need for more complete studies, including the fruit production phase. It is important to note that this work has provided

various methodologies for diagnosing the quality of substrates and the growth capacity of A. brasiliensis , with

a view to obtaining maximum reproductive biomass.

Finally, the results obtained provided technical and scientific tools for monitoring the growth of A.

brasiliensis, which is currently in the spotlight due to its commercial interest, both in traditional cultivation

substrates and using alternative strategies.

## BIBLIOGRAPHICAL REFERENCES


AGENDA 21, Department of Agriculture and Environment - SAMA. Government of Joinvile - SC. 266p. 1997.


AN Jaraguá. Available at http://www.an.com.br. Accessed on: 01 June 2003.


Attenuated Total Reflectance Fourier Spectroscopy (ATR/ FTIR) Available at www.micromemanalytical.com. Accessed on: 07 Sep 2004.


ANDER, P. & ERIKSSON, K. E. The importance of phenol oxidase activity in lignin degradation by the white rot fungus Sporotrichum pulverulentum . **Archives of Microbiology,** 109, 1-8, 1976.


BAIYLEY, M. J., BYELI, P., POUTANEN, K. Interlaboratory testing of methods for xylanase activity, **Journal of Biotechnology,** v. 23, n. 3, p. 257-270, 1992.


BARAJAS -ACEVES, M.; HASSAN, M.; TINOCO, R. & VAZQUEZ-DUHALT, R. Effect of pollutants on the ergosterol content as indicator of fungal biomass. **Journal Microbiology Methods, 50,** 227 - 236, 2002.


BOLLAG, J.-M. & LEONOWICZ, A. Comparative studies of extracellular fungal laccases. **Applied Environmental Microbiology.** 48, 849-854, 1984.


BRAGA, G. C.; EIRA, A. F.; CELSO, P. G. & COLAUTO, N. B. **Manual do cultivo de** *Agaricus blazei* Murril "Cogumelo-do-Sol". FEPAF - Fundação de Estudos e Pesquisas Agrícolas e Florestais, Botucatu, SP. 44p. 1998.


BRAGA, G. C. & EIRA, A. F. Manual de cultivo de **"champignon". Unesp. Foundation for** agricultural and forestry studies and research. Botucatu-SP.1998.45p.


BRADFORD, M. M. A rapid and sensitive method for the quantification of microgram quantities of protein utilizing the principle of protein dye binding. **Anal. Biochemistry.**p. 248-254, 1976.

BUSWELL, J. A.; CAI, Y. J.; CHANG, S. T.; PEBERDY, J. F.; FU, S. Y. and YU, H.-S. Lignocellulotytic enzyme profiles of edible mushroom fungi. **World Journal of Microbiology and Biotechnology**, v.12, p. 537-542, 1996.

BONNEN, A.M. ANTON, L.H. & ORTH, A.B. Lignin-degrading enzymes of the commercial button mushroom, Agaricus bisporus. **Applied and Environmental Microbiology**, 60. 960-965, 1994.

BUSWELL, J. A.; CAI, Y.J.; CHANG, S.T. Fungal- and substrate-associated factors affecting the ability of individual mushroom species to utilise different lignocellulosic growth substrates. Chapter 15 in: CHANG, S.T.; BUSWELL, J.A.; CHIU, S. **Mushroom Biology and Mushroom Products**. The Chinese university press. Hong Kong. 370p. 1993.

CAI, Y. J.; CHAPMAN, S.J.; BUSWELL. J.A.; CHA NG. S.T. Production and distribution of endoglucanase, cellobiohydrolase, and â-glucosidase components of the cellulolytic system of Volvariella volvacea, the edible straw mushroom. **Applied and Environmental Microbiology**, 553-559, 1999.

CAI, J. Y.; BUSWELL, J. A. & CHANG, S. T. â-glucosidase components of the cellulolytic system of the edible straw mushroom, Volvariella volvacea. **Enzyme and Microbial Technology**, 22: 122-129, 1998.

CAMPBELL, A. C. & SLEE, R. W. Commercial cultivation of Shiitake in Taiwan and in Japan. **Mushroom & Tropics**, 7, 127-137, 1987.

CHANG, S.T.; BUSWELL, J.A.; CHIU, S. Mushroom biology and mushroom products. The Chinese university press. Hong Kong. 370p. 1993.

CHEN, Y.; CHEFETZ, B.; ROSARIO, R.; VAN HEEMST, J.D.H.; ROMAINE, C.P. & HATCHER, P.G. Chemical nature and composition of compost during mushroom growth. **Compost and Science & Utilisation**, v. 8, No. 4, p. 347 - 359, 2000.

DING, S.; GE, W.; BUSWELL, J.A.; Endoglucanase I from edible straw mushroom, Volvariella volvacea. Purification, characterisation, cloning and expression. **Euro Journal Biochemistry**, 268. 5687-5695, 2001.

DURRANT, A. J.; WOOD, D. A. & CAIN, R. B. Lignocellulose biodegradation by Agaricus bisporus during solid state fermentation. **Journal of Genera Microbiology**, 137, 751-755, 1991.

EIRA, A. F. Cultivation of medicinal mushrooms. Ed. Aprenda fácil. São Paulo. 395p. 2003.

EIRA, A. F. Mushroom cultivation (composting, management and environment). Proceedings of the **III Itinerant Plant Health Meeting of the Biological Institute**, Mogi das Cruzes, SP. 83-95 p. 2000.

FAIX, O. Investigation of lignin polymer models (DHP's) by FTIR spectroscopy. **Holzforschung**, (5) 40. 273-280, 1986.

FAIX, O.; MOZUCH, M. D. & KIRK, T. K. Degradation of gymnosperm (Guaiacyl) vs. Angiosperm (Syringyl/Guaiacyl) lignins by Phanerochaete chrysosporium. **Holzforschung,** 39: 203-208, 1985.

FERREIRA, J. E. F. Production of mushrooms. Livraria e editora agropecuária Ltda, Guaíba - RS. 136p. 1998.

FTIR Spectroscopy. Available at www.nelsonlabs.com , Accessed on: 20 Dec 2004.

GALLIANO, H.; GAS, G.; SERIS, J.L. & BOUDET, A.M. Lignin degradation by Rigidopuros lignosus involves synergistic action of two oxidising enzymes: Mn peroxidase and laccase. **Enzyme and Microbial Technology,** 13, 478-482, 1991.

GIVEN, P. H.; SPACKMAN, W.; PAINTER, P.C.; RHOADS, C.A.; RYAN, N. J.; ALEMANY, L. & PUGMIRE, R. J. The fate of cellulose and lignin in peats: an exploratory study of the input to coalification. **Organic Geochemistry,** 6: 399 - 407, 1984.

GERRITS, J.P.G. Nutrition and compost. In: **The cultivation of mushrooms.** First English edition, 1988, 29-72.

HATAKKA, A. Lignin-modifying enzymes from selected White-rot fungi: production and role in lignin degradation. **FEMS Microbiol Rev.,** 13, 125-135, 1994.

HIROTANI, M.; SAI, K.; ASAMI, K.; YOSHIHISA, A.; AND TAKAFUMI, Y.
Biosynthetic studies on blazeispirane and protoblazeispirane derivatives from the cultured mycelia of the fungus *Agaricus blazei* . **Tetrahedron,** 58: 10251- 10257, 2002.

IIYAMA, K.; STONE, B.A. & MACAULEY, B.J. Compositional changes in compost during composting and growth of *Agaricus bisporus.* **Applied Environmental Microbiology,** 60, 1538-1546, 1994.

INBAR, Y.; CHEN, Y. & HADAR, Y. Solid-state Carbon-13 Nuclear magnetic resonance and infrared spectroscopy of composted organic matter. **Soil Science Society American Journal,** 53, 1695-1701, 1989.

KAKEZAWA, M.; MINURA, A.; TAKAHARA,Y. Application of two-step composting process to rice straw compost. **Soil Science and Plant Nutrition,** 38(1), 43-50, 1990.

KEREM, Z. & HADAR, Y. Effect of manganese on lignin degradation by Pleurotus ostreatus during solid-state fermentation. 59:12, 4115-4120, 1993.

KEREM, Z.; FRIESEM, D. & HADAR, Y. Lignocellulose degradation during solid- state fermentation: Plourotus ostreatus versus Phanerochaete crysosporium. **Applied and environmental microbiology,** 58 (4): 1121-1127, 1992.

KOPYTOWSKI FILHO, J. C/N ratio and proportion of nitrogen sources in the productivity of Agaricus blazeii Murrill and calorific value of compost. 2002. 96p. (Master's dissertation). UNESP, Botucatu, SP.

KUES.V & LIU. Y. Fruting body production in basidiomycetes. **Applied Microbiology and Biotechnology**, 54: 141-156, 2000.

KUWAHARA, M.; GLENN, J. K.; MORGAN, M. A.; GOLD, M. H. Separation and characterisation of two extracellular H2O2-dependent oxidases from ligninolytic cultires of Phanerochaete chrysosporium . **FEBS Letters**, v. 169, n. 2, p.247- 250, 1984.

LAW, W.M ;LAU, W.N.; LO, K.L.; WAI, L.M. & CHIU, S.W. Removal of biocide pentachlorophenol in water system by the spent mushroom compost of Pleurotus pulmonarios. **Chemosphere**, 52. 1531-1537, 2003.

MACAULEY, B. J. & BETHEL, M. Ergosterol as an indicator of biomass of Agaricus brunnescens mycelium grown under varying cultural conditions. **Fourth International Mycological Congress.** Regenburg, Germany. Abstracts: 331.1990

MANDELS, M.; ANDREOTTI, R.; ROCHE, C. Measurement of saccharifying cellulase. **Biotechnology Bioengineering**, v. 6, p. 21-33, 1976.

MATCHAM, S.E; JORDAN, B.R. & WOOD, D.A. Estimation of fungal biomass in a solid substrate by tree independent methods. **Applied Microbiology and Biotechnology**, 21: 108-112, 1985.

MAYER, A.M. & STAPLES, R.C. Laccase: new functions for an old enzyme. **Phytochemistry**. 60, 551-565, 2002.

MILLER, G.L. Use of dinitrosalicylic acid reagent for determination of sugar reducing. Analytical Chemistry, **31**, 424-426, 1959.

MOL, P. C. Hyphal wall elongation during expansion growth of the common mushroom (Agaricus bisporus). **Rijksuniversiteit Groningen**, 117p, 1989.

MOLENA, O. Modern mushroom cultivation. Ed. Nobel. 170p. 1986.

MOLIN, P.; P. GERVAIS, J. P. LEMIÃRE AND T. DAVET. Direction of hyphal growth: a relevant parameter in the development of filamentous fungi. **Research in Microbiology** v. 143,8. 777-784, 1992.

MUDGETT, R. E. Solid-state fermentation. Chapter 7 in: DEMAIN. A. L. & SOLOMON, N. A. **Manual of industrial microbiology and biotechnology**, Washington. 486p. 1986.

NASCIMENTO, J. S & EIRA, A. F. Occurrence of the False Truffle (Diehliomyces microsporus Gilkey) and demage on the Himematsutake Medicinal Mushroom (Agaricus brasiliensis S. Wasser et al.). **International Journal of Medicinal Mushroom**, 5, 87-94, 2003.

NEVES, M. A. Ecological, physiological and genetic characterisation of Agaricus blazei Murrill using strains from different mushroom production companies. 2000. 69p. (Master's dissertation). UFSC, Florianópolis, SC.

NOBLE, R.; GROGAN, H.M. & ELLIOT, T. Variation in morphology, growth and fructification of isolates in the Agaricus subfloccosus complex. **Mycologycal Research**, 99 (12): 1453-1461, 1995.

OKEKE, B.C.; PATERSON, A.; SMITH, J.E. & WATSON-CRAIK, I.A. The relationship between phenol oxidase activity, soluble protein and ergosterol with growth of Lentinus species in oak sawdust logs. **Applied Microbiology and Biotechnology 41**, 28-31, 1994.

PANDEY, K. K. & PITMAN, A. J. FTIR studies of the changes in wood chemistry following decay by brown-rot and white-rot fungi. **International Biodeterioration & Biodegradation.** 52, 151-160, 2003.

PIET, W.J; DE GROOT; BASTEN, D.E.J.W; SONNEMBERG, A.S.M; VAN GRIENSVEN L.J.L.D; VISSER, J. & SCHAAP, P.J. An Endo-1,4-â -xylanase- encoding gene from Agaricus bisporus is regulated by compost- specific factors. **Journal Molecular Biology**, 277, 273-284, 1998.

REID, I. D. Solid state fermentation for biological delignification. **Enzyme and Microbial Technology,** 11: 786-803, 1989.

RUTTIMANN, C.; SCHWEMBER, L.; SALAS, D.; CULLEN, D. & VICUNA. Lignolytic enzymes of the white-rot basidiomycetes Phlebia brevispora and Ceriporiopsis subvermispora. **Biotechnology Applied Biochemistry,** 16, 64-76, 1993.

SÁNCHES-VÁZQUEZ, J. E. Research update on growing Portobello and A. blazei on non-composted substrates. **IX Annual Specialty Mushroom Workshop,** Pennsylvania State University, PA. 1999.

SANTOS, V. M. C. S. Contribution to the study of the production of Pleurotus spp. Master's thesis. UFSC. Florianópolis, SC. 128p. 2000.

SAVOIE, J-M; BRUNEAU, D. & MAMOUN, M. Resource allocation ability of wild isolates of Agaricus bisporus on conventional mushroom compost. **FEMS Microbiol ecology** 21: 285-292, 1996.

SEITZ, M. L.; SUER, D. B.; BURROUGHS, H.E; HUBBARD, J.D. Ergosterol as a measure of fungal growth. **Phytopathology,** 69: 1202-1203, 1979.

SHARMA, H. S. S. & KILPATRICK, M. Use of Near-Infrared Spectroscopy to predict potential mushroom (Agaricus bisporus) yield of phase II compost. **Applied Spectroscopy,** v 54: 1, 44-47, 2000.

SCHISLER, L.C.; Biochemical and mycological aspects of mushroom composting. **Penn State Handbook for Commercial Mushroom Growers,** p. 3-14, 1982.

SOARES, C. H. L. & DÚRAN, N. Biodegradation of chorolignin and lignin-like compounds contained in E-

Pulp bleaching effluent by fungal treatment. **Applied Biochemistry and Biotechnology.** 95, 135-149, 2001.

SOARES, C. H. L. & DÚRAN, N. FT-IR spectroscopy of E1 -pulp blanching effluent degraded by basidiomycete fungi. In: Proceedings of Fourth Brazilian Symposium on the Chemistry of Lignins and Other Wood Components. 1-4, V.5, 1995.

STAMETS, P. Growing gourmet and medicinal mushrooms. 3ª ed. Ten Speed Press. Berkeley, California. 592p. 2000.

STRAATSMA, G.; DI LENA, G.; OLIJNSMA, T.W.; OP THE CAMP, H. J. M.; & VAN GRIENSVEN, L. J. L. D. Laboratory media for measuring growth parameters of Agaricus bisporus mycelium as influenced by Scytalidium thermophilum . **Cultivated Mushroom Research**, 1: 1-6, 1993.

STURION, G. L. Use of banana leaves as a substrate for growing Pleurotus spp. mushrooms Master's thesis. USP. Piracicaba, SP. 147p. 1994.

TAKATU, T.; KIMURA. Y; OKUDA, H. Isolation of an antitumour compound from Agaricus blazei Murill and its m echanism of action. **Journal Nutrition**, 131(5): 1409-1413, 2001.

TEDESCO, M. J.; GIANELLO, C.; BISSANI, C. A.; BOHNGN, H. & VOLKWEISS, S.J. Analysis of soil, plants and other materials. 2nd ed. Department of Soils, Faculty of Agronomy, Rio Grande do Sul, Porto Alegre, 174p. 1985.

TIEN, M. & KIRK, T. K. Lignin-degrading enzyme from Phanerochaete chrysosporium purification, characterisation, and catalytic properties of a unique H2O2-requiring oxygenase. **Proc. Natl. Acadademy Science**, v.81, p.2280-2284, 1984.

TILL, O. Champignonkultur auf sterilisiertm naehrsubstrat und die wiederverwendung von adggetragenem compost. **Mushroom Science**, 5: 127 -133, 1962.

TONINI, R. C. G. Study of the feasibility of using the palm heart sheath (Euterpe edulis) as a substrate for growing Lentinula edodes (Beck.) Pegler. Master's dissertation. FURB. Blumenau. Santa Catarina. 153p. 2004.

TOUMELA, M.; VILKMAN, M.; HATAKKA, A. & ITÃVAARA. Biodegradation of lignin in a compost environment: a review. **Bioresource Technology.** 72, 169-183, 2000.

VÁLAZQUEZ-CEDENO, M.A.; MATA, G.; SAVOIE, J.M. Waste reducing cultivation of Pleurotus ostreatus and Pleurotus pulmonarius on coffe pulpe changes in the production of some lignocellulolytic enzymes. **Word Journal of Microbiology and Biotechnology,** v.18, n.3. 201-207 (7), 2002.

VILLAS -BÔAS, S.G., ESPOSITO, E. AND MITCHELL, D.A. Microbial conversion of lignocellulosic residues for production of animal feeds. **Animal Feed Science Technology,** 98:1-12, 2002.

YANG et al. Lignocellulose degradation during growth of fungus Pholiota nameko in: CHANG, S.T.; BUSWELL, J.A.; CHIU, S. Mushroom biology and mushroom products. The Chinese university press. Hong Kong. 370p. 1993.

WANG, Z.; CHEN, T.; GAO, Y.; BREUIL, C. & HIRATSUKA, Y. Biological degradation of resin acids in wood chips by wood-inhabiting fungi. **Applied and Environmental Microbiology,** 61(1): 222-225, 1995.

WASSER, S. P.; DIDUKH, M.Y.; AMAZONAS, M. A. L.; NEVO, E.; STAMETS. P.; EIRA, A. F. Is a Widely Cultivated Culi nary-Medicinal Royal Sun Agaricus (the Himematsutake Mushroom) Indeed Agaricus blazei Murrill? **International Journal of Medicinal Mushroom,** v.4, p.297-90, 2002.

WOOD, D. A. & GOODENOUGH, P. W. Fruiting of Agaricus bisporus: changes in extracellular enzym e activities during growth and fruiting. **Arch. Microbiology,** 114, 161-165, 1977.

WOOD, D. A. Inactivation of extracellular laccase during fruiting of Agaricus bisporus. **Journal of General Microbiology,** 117, 339-345, 1980.